# LABORATORY EXERCISES IN ENVIRONMENTAL GEOLOGY

# LABORATORY EXERCISES IN ENVIRONMENTAL GEOLOGY

HARVEY BLATT
*University of Oklahoma*

WCB Wm. C. Brown Publishers
Dubuque, Iowa • Melbourne, Australia • Oxford, England

**Book Team**

Editor *Jeffrey L. Hahn*
Production Editor *Kay J. Brimeyer*
Designer *Eric Engelby*
Art Editor *Kathleen M. Huinker*
Photo Editor *Diane S. Saeugling*

**Wm. C. Brown Publishers**
A Division of Wm. C. Brown Communications, Inc.

Vice President and General Manager *Beverly Kolz*
Vice President, Publisher *Kevin Kane*
Vice President, Publisher *Earl McPeek*
Vice President, Director of Sales and Marketing *Virginia S. Moffet*
Marketing Manager *Christopher T. Johnson*
Advertising Manager *Janelle Keeffer*
Director of Production *Colleen A. Yonda*
Publishing Services Manager *Karen J. Slaght*

**Wm. C. Brown Communications, Inc.**

President and Chief Executive Officer *G. Franklin Lewis*
Corporate Senior Vice President, President of WCB Manufacturing *Roger Meyer*
Corporate Senior Vice President and Chief Financial Officer *Robert Chesterman*

Cover photo obtained from the Hong Kong Government, Geotechnical Engineering Office

Copyedited by Katherine Stevenson



A Times Mirror Company

Library of Congress Catalog Card Number: 93–70603

ISBN 0–697–17071–3



Printed in the United States of America by Wm. C. Brown Communications, Inc., 2460 Kerper Boulevard, Dubuque, IA 52001

10 9 8 7 6 5 4 3 2 1

***Art Credits***

**JAK Graphics, Ltd:**
1.1, 3.3a, 5.1, 5.4, 5.5, 6.3, 6.4, 7.1a&b, 7.3, 7.4, 7.5, 8.1, 8.2, 8.3, 8.4a, 9.3, 9.4, 9.5, 10.1a, 10.3, 11.1, 11.2, 11.3a&b, 11.4a&b, 11.5, 11.6, 12.1, 12.4, 12.6a&b, 14.1, 14.2, 14.5a&b, 14.8a&b, 15.1a, 15.2, 15.4, 16.2, 17.2a&b, 18.2, 18.3, 19.1, 19.3, 19.5, 20.1, 20.2, TA 21.1, 22.2, 22.3, 22.4a&b, 22.5, 23.2, 24.2a-c, 24.3.

***Photo Credits***

**Chapter 1**
**p. 9** (Lt. & Rt.): © WCB, Robert Rutford, & James Zumberge/James Carter, Photographer; **p. 10, 11** (Bottom Left): Photo by C.C. Plummer; **p. 11** (Bottom Left), **p. 12** (Top Left): © WCB, Robert Rutford, & James Zumberge/James Carter, Photographer; **p. 11:** © Wm.C.Brown Publishers/Photograph by Bob Coyle.

**Chapter 2**
**p. 17** (Top Left, Bottom Left, Top Right, Middle Right): © WCB, Robert Rutford, & James Zumberge/James Carter, Photographer; **p. 18** (Top Left, Top Right, Bottom Right): © WCB, Robert Rutford, & James Zumberge/James Carter, Photography; **p. 18:** © Omni Resources/ Wm. C. Brown Communications.

**Chapter 3**
**Fig. 3.4C:** © Omni Resources/Wm. C. Brown Communications; **3.4D:** © WCB, Robert Rutford, & James Zumberge/James Carter, Photographer.

**Chapter 4**
**Fig. 4.1c, 4.1d, 4.1e, 4.1f:** © WCB, Robert Rutford, & James Zumberge/James Carter Photographer; **4.1G:** Photo by C.C. Plummer.

**Chapter 6**
**Fig. 6.5A:** U.S. Geological Survey.

**Chapter 8**
**Fig. 8.4a:** Courtesy from Eric Fantaine; **8.4b:** NOAA Satellite Research Laboratory, courtesy Larry Stowe and Robert Carey.

**Chapter 9**
**Fig. 9.6:** Hamblin/Howard: Exercises in Physical Geology, 7/e, **8.8 p.85** MacMillan Publishing.

**Chapter 10**
**Fig. 10.6:** Geological Library/University of Oklahoma.

**Chapter 11**
**Fig. 11.7:** U.S. Geological Survey.

**Chapter 12**
**Fig. 12.3:** "Cohesion in Undisturbed Sensitive Clay", Geotechnique Vol. XIII 1963, **Fig. 1 page 134** (a); **12.5:** Photo obtained from the Hong Kong Government, Geotechnical Control Office; **12.7:** U.S. Geological Survey.

**Chapter 13**
**Fig. 13.1A:** Blackwell Scientific Publishers; **13.1B:** AAPG Memoir 38, 1984 American Association of Petroleum Geologists **1A p.208; 13.2:** SEPM; **13.3:** U.S. Bureau of Reclamation; **13.4:** Arthur N. Palmer GSA Bulletin, Jan. 1991, Vol.103-1 (cover photo) GSA.

**Chapter 14**
**Fig. 14.3:** © Harvey Blatt; **14.4:** N.H. Darton U.S. Geological Survey Water-Supply Paper 227; photo courtesy R.E. Fidler.; **14.7:** Photo courtesy of George Remaine, Orlando Sentinel Star.

**Chapter 15**
**Fig. 15.1B:** U.S. Geological Survey; **15.3:** The Department of Oil Properties, City of Long Beach.

**Chapter 17**
**Fig. 17.1A:** U.S. Department of Agriculture; **17.1B:** U.S. Department of Agriculture; **17.3:** Indiana Geological Survey; **17.4A-B:** (a) Metropolitan Museum of Art (b) New York City Parks Photo Archive.

**Chapter 18**
**Fig. 18.1:** Petrosystems.

**Chapter 23**
**Fig. 23.3:** NASA and NOAA Landsat, courtesy Barrett N. Rock and James Vogelmann, University of New Hampshire; David Zlotek, Cirrus Technology; and Hanan Kadro, University of Freiburg/FRG.

contents

The past decade has witnessed an explosion of interest in the relationship between geology and the everyday concerns of the average citizen. But why should someone who is not a professional geologist care about planet Earth? So what if the Earth's surface is composed of plates, rather than an unbroken, homogeneous skin? Why should anyone care whether or not a rock layer contains oriented clay minerals? Will the fact that calcium carbonate is more soluble than quartz affect *your* life in any meaningful way?

The answers to these questions reveal that the average citizen can indeed be affected by such concerns. The fractured, platy character of the Earth's crust, for example, affects the cost of house insurance in California and elsewhere. The presence of oriented clay minerals predisposes inclined rock layers to slide downhill in many parts of the United States. The difference in solubility between calcium carbonate and quartz causes some Florida homes to collapse into huge pits.

The relationship between geology and short-term human concerns (periods of no more than a few hundred years) is termed *environmental geology*. The purpose of this laboratory manual is to provide examples of how rocks and minerals exposed at the Earth's surface and geological processes affect the natural environment.

At most schools, environmental geology is taught at the first-year level, commonly as an alternative to a standard Physical Geology course, and is designed for students who need to fulfill a college science requirement. The course also serves as an elective for students interested in the environment. Because of the dual function of this course, students in the class often have quite varied science backgrounds. Some have had a previous university science course, while others have had none, and might even have avoided science and mathematics in high school. To help accommodate the needs of both groups, the exercises in this laboratory manual typically contain more questions than can be answered within a standard two- to three-hour laboratory class, and the questions vary in difficulty. In addition, each exercise contains a mixture of hands-on and thought-provoking questions that deal with social, ethical, or political issues of environmental relevance. The instructor can use whichever questions seem most appropriate.

The manual also contains more than enough exercises for the normal 14 labs per semester, so instructors can choose the exercises most suitable for their geographic area. For example, problems of floods and coastal erosion seem more immediate in Louisiana than in Minnesota.

This manual is designed to be used in conjunction with textbooks on physical geology and environmental geology. For this reason I have avoided the lengthy theoretical discussions normally included in those texts. I have, however, described some of the basic principles that underlie each laboratory exercise.

These principles cannot be discussed without reference to size and distance. How far is a house from an unstable hillside? Is the hill slope steep or gentle? Scientists usually use a metric scale in answering such questions; engineers and other professionals use the familiar, nonmetric scale of feet and inches. Scale problems also arise with temperature measurements. Scientists use the centigrade or Kelvin scale rather than the Fahrenheit scale more familiar to American students. To help students gain familiarity with various systems, this manual uses different scales in different exercises. Students also need to learn how to convert easily from one set of units to another; therefore, numerous conversion factors are included in Appendix A.

Laboratory time is perhaps too often envisioned as time spent inside a building, despite the fact that geology is largely an outdoor science. Problems in geology occur where there are rocks. I suggest that several laboratory periods be devoted to field excursions to geologically and environmentally important sites near the university. Examples might include slumps, landslides, building damage resulting from swelling clays, sanitary landfills, coal mines, abandoned mines and their waste piles, houses built on river floodplains, or any of the many other scenes of interactions between humans and the Earth.

During the past 20 years, the increasing importance of environmental problems has led to the publication of many books that are important references for environmental concerns. Books that deal with geological influences on the environment include the following:

Coates, D. R., 1981, *Environmental Geology.* New York, John Wiley & Sons, 701 pp.

Coates, D. R., 1985, *Geology and Society.* New York, Chapman and Hall, 406 pp.

Costa, J. E., and Baker, V. R., 1981, *Surficial Geology.* New York, John Wiley & Sons, 498 pp.

Dennen, W. H., and Moore, B. R., 1986, *Geology and Engineering.* Dubuque, Iowa, Wm. C. Brown, 378 pp.

Garrels, R. M., Mackenzie, F. T., and Hunt, C., 1975, *Chemical Cycles and the Global Environment: Assessing Human Influences.* Los Altos, California, William Kaufmann, 206 pp.

Goudie, A., 1990, *The Human Impact on the Natural Environment,* 3rd ed. Cambridge, Massachusetts, MIT Press, 388 pp.

Griggs, G. B., and Gilchrist, J. A., 1983, *Geologic Hazards, Resources, and Environmental Planning,* 2nd ed. Belmont, California, Wadsworth, 502 pp.

Keller, E. A., 1992, *Environmental Geology,* 6th ed. Columbus, Ohio, Merrill, 521 pp.

Legget, R. F., 1973, *Cities and Geology.* New York, McGraw-Hill, 624 pp.

Leveson, D., 1980, *Geology and the Urban Environment.* New York, Oxford University Press, 386 pp.

Lundgren, L., 1986, *Environmental Geology.* New York, Prentice-Hall, 576 pp.

Montgomery, C. W., 1992, *Environmental Geology,* 3rd ed. Dubuque, Iowa, Wm. C. Brown, 465 pp.

Ward, K. (ed.), 1989, *Great Disasters.* Pleasantville, New York, Reader's Digest Association, 320 pp.

I hope students will finish this laboratory manual with a better understanding and appreciation of the ground beneath their feet. I encourage both faculty and student users of this manual to send suggestions for improving any of the exercises to either the author or the publisher.

I wish to extend a special thanks to the following people for their thoughtful comments and suggestions in prepublication reviews:

John R. Huntsman
*University of North Carolina at Wilmington*

Rob Sternberg
*Franklin and Marshall College*

Mary P. Anderson
*University of Wisconsin–Madison*

Paul D. Nelson
*St. Louis Community College at Meramec*

George P. Merk
*Michigan State University*

James R. Lauffer
*Bloomsburg University*

William D. Nesse
*University of Northern Colorado*

Jim Constantopoulos
*Eastern New Mexico University*

David L. Ozsvath
*University of Wisconsin–Stevens Point*

Their contributions were enlightening, challenging, and encouraging throughout the development process.

# exercise ONE

# Minerals

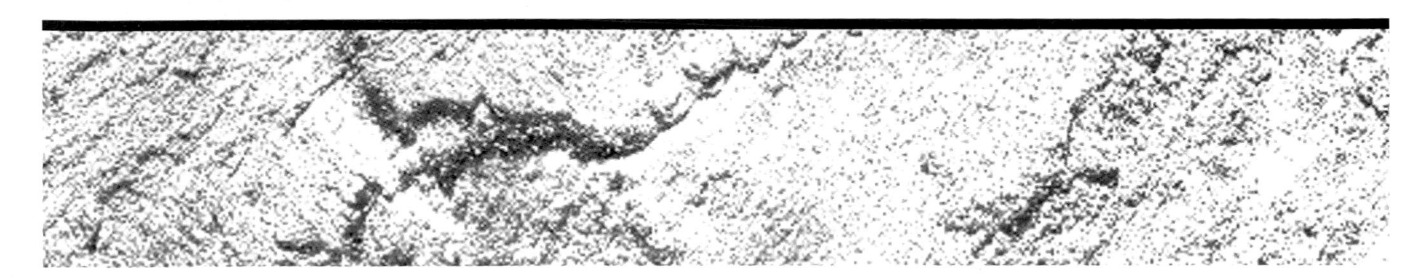

Although our planet is composed of 88 chemical elements, they are not present in equal amounts. Eight elements form 98.5% by weight of the *crust*—the upper 30 miles under the land surface—while the other 80 elements total only 1.5%. The abundant elements are

| Element | Weight (%) |
|---|---|
| Oxygen | 46.6 |
| Silicon | 27.7 |
| Aluminum | 8.1 |
| Iron | 5.0 |
| Calcium | 3.6 |
| Sodium | 2.8 |
| Potassium | 2.6 |
| Magnesium | 2.1 |
| All others | 1.5 |

The composition and relative abundance of minerals reflect the chemical composition of the crust. The most abundant minerals are composed largely of oxygen, silicon, and aluminum.

Also noteworthy is the fact that most of the economically important elements (erroneously called "minerals" on breakfast-cereal boxes) are not among the eight most abundant elements. For example, titanium makes up only 0.57% of the crust of the Earth, manganese is only 0.09%, chromium is only 0.01%, and other elements such as nickel, copper, and lead are even less abundant. Concentrations of most of these elements are uncommon—and are becoming even less so as our industrial civilization expands. Substitutes for most of them have yet to be found or synthesized.

Elements tend to combine into larger groupings because of their electronic structures. The change in electron distribution that results when elements combine determines the physical and chemical characteristics of the materials (e.g., minerals) produced. Unfortunately, however, the new chemical properties cannot be predicted from the properties of the individual, uncombined elements. For example, at room temperature sodium (Na) is a metal and chlorine (Cl) is a gas, but when the two combine as sodium chloride (NaCl, halite), they produce a solid—ordinary table salt. The properties of table salt (including very high solubility in water or in steak juice) are determined by the distribution of electrons in this sodium chloride aggregate, just as the properties of the uncombined sodium (metallic appearance, high melting temperature) and chlorine (irritating odor, greenish yellow color) were determined by *their* electronic structures.

Just as the properties of solid NaCl differ from those of uncombined sodium and chlorine, so do they also differ from the properties of sodium and chloride ions (charged atoms) dissolved in water. The "salty" taste sodium ions create in water (or saliva) is well known. The arrangement of electrons around atomic nuclei underlies the physical and chemical properties of all materials.

Two properties of great importance to environmental scientists studying minerals are hardness and solubility. Quartz ($SiO_2$) is very hard and insoluble in water; calcite ($CaCO_3$) is soft and moderately soluble in water; halite (NaCl) is very soft and very soluble in water. The importance of these chemical properties to drinking-water quality and to building construction in humid climates is obvious.

## *Important Properties of Minerals*

A mineral is a naturally occurring, inorganic solid with a regular, periodic internal structure and a fairly definite chemical composition (table 1.1). Because of this fixed internal structure and chemical composition, the physical and chemical properties of a mineral are constant and can be used to identify it. The following properties are those most useful in mineral identification:

1. *Hardness,* defined as the ability of a mineral to resist abrasion, is determined by scratching the mineral with an object of known hardness. Harder minerals scratch softer ones. Geologists use a hardness scale devised in 1824 by a German mineralogist, Friedrich Mohs, and known as the *Mohs hardness scale* (table 1.2). Surface alteration can decrease hardness; therefore, hardness must be determined on a fresh mineral surface.
2. *Cleavage.* The strength of chemical bonds in a mineral differs for different pairs of elements. Because of this, and because the elements occur in fixed positions in its crystal structure, a mineral can have some planar surfaces across which bonding is weaker. When hit, the mineral tends to break along these weaker planes, which are called *cleavage surfaces* (figure 1.1). Minerals can have one, two, three, four, or six different cleavage directions, and these can be diagnostic for the mineral. For example, micas have one cleavage direction and occur as sheets. In a highly micaceous rock, slippage and slope failure tend to parallel oriented groups of mica flakes.

   Some minerals do not show cleavage, either because their cleavage surfaces are poorly developed or because their chemical bonds are nearly equal in all directions. Quartz is a mineral with no obvious cleavage.

   Cleavage faces should not be confused with crystal faces. Cleavage faces are planar surfaces of preferential breakage that reflect planes of weakness in a crystal structure. Crystal faces, in contrast, are external planar surfaces that form as a mineral grows from a solution; they reflect the geometry of the internal structure. Only rarely is the shape of cleavage fragments of a mineral the same as the shape of its fully formed crystals. Crystals normally are bounded by many more surfaces intersecting at different angles than are cleavage fragments. Among the few examples of cleavage fragments identical to fully formed crystals are halite cubes, galena cubes, and dolomite rhombohedra.

TABLE 1.1

Common Rock-Forming Minerals

| **Abundant** | |
|---|---|
| **Mineral** | **Chemical Composition** |
| Quartz | $SiO_2$ |
| Orthoclase feldspar | $KAlSi_3O_8$ |
| Plagioclase feldspar | $NaAlSi_3O_8$ to $CaAl_2Si_2O_8$ |
| Biotite mica | hydrous K, Fe, Mg, Al silicate |
| Muscovite mica | $KAl_2AlSi_3O_{10}(OH)_2$ |
| Hornblende | Ca, Na, Mg, Fe, Al silicate |
| Augite | Ca, Mg, Fe, Al silicate |
| Olivine | $(Mg, Fe)_2SiO_4$ |
| Chlorite | hydrous Mg, Fe, Al silicate |
| Illite clay | hydrous K, Al, Fe, Mg silicate |
| Montmorillonite clay | hydrous Na, Ca, Al, Fe, Mg silicate |
| Kaolinite clay | $Al_2Si_2O_5(OH)_4$ |
| Calcite | $CaCO_3$ |
| Dolomite | $CaMg(CO_3)_2$ |
| Gypsum | $CaSO_4 \cdot 2H_2O$ |
| Halite | NaCl |
| Hematite | $Fe_2O_3$ |
| **Less Abundant but Still Common** | |
| **Mineral** | **Chemical Composition** |
| Garnet | Fe, Mg, Ca, Al silicate |
| Kyanite | $Al_2SiO_5$ |
| Sillimanite | $Al_2SiO_5$ |
| Staurolite | Fe, Mg, Al silicate |
| Epidote | Ca, Fe, Al silicate |
| Magnetite | $Fe_3O_4$ |
| Ilmenite | $FeTiO_3$ |
| Pyrite | $FeS_2$ |
| Graphite | C |

TABLE 1.2

Mohs Hardness Scale

| Relative Hardness | Index Mineral | Common Objects |
|---|---|---|
| 10 | Diamond | |
| 9 | Corundum | |
| 8 | Topaz | |
| 7 | Quartz | |
| | | Steel file—6.5 |
| 6 | Orthoclase | |
| | | Glass, knife, nail—5.5 |
| 5 | Apatite | |
| 4 | Fluorite | |
| | | Copper penny—3.0 |
| 3 | Calcite | |
| | | Fingernail—2.5 |
| 2 | Gypsum | |
| 1 | Talc | |

**Figure 1.1**

Cleavage patterns of minerals. Few common minerals have more than three cleavage directions.

| Number of cleavage directions | | Shape | Sketch |
|---|---|---|---|
| 0 | No cleavage, only fracture | Irregular masses (quartz) |  |
| 1 | | Flat sheets (micas) |  |
| 2 at 90° | | Elongated form with rectangular cross-section (prism: spodumene) | 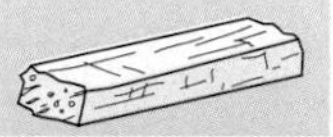 |
| 2 not at 90° | | Elongated form with parallelogram cross-section (prism: hornblende) | 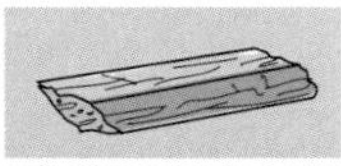 |
| 3 at 90° | | Cube (halite) |  |
| 3 not at 90° | | Rhombohedron (calcite) |  |
| 4 | | Octahedron (fluorite) | 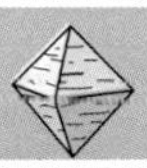 |
| 6 | | Dodecahedron (sphalerite) |  |

3. *Color.* Many of the abundant or common rock-forming minerals have distinctive colors that are useful for identification. Some minerals (particularly quartz, fluorite, and calcite) can occur in a wide variety of colors, but one color is the most common. Colors generally result from the presence of impurities that selectively absorb wavelengths of light entering the mineral. Those wavelengths that are not absorbed give the mineral its color.
4. *Streak,* the color of the mineral powder, is determined by powdering a sample, usually by scratching it across a piece of unglazed porcelain (hardness 7). The mineral must be softer than the porcelain, or it will scratch (powder) the porcelain rather than being powdered itself. Most nonmetallic minerals have a white or colorless streak; hence streak is not a helpful diagnostic tool for the abundant minerals, almost all of which are nonmetallic. Streak is more helpful for identifying metallic minerals, many of which are of great economic importance.
5. *Luster* is the appearance of a fresh mineral surface in reflected light. A mineral that appears metallic is said to have a *metallic* luster. Nonmetallic mineral surfaces can be *vitreous* (glassy luster), *resinous, pearly, silky, dull,* or *earthy.* As examples, quartz is vitreous, sphalerite is resinous, talc is pearly, satin spar gypsum is silky, and microcrystalline hematite is dull or earthy.
6. *Specific gravity* is the ratio between the weight of a mineral and the weight of an equal volume of water. Most minerals have specific gravities of between 2.6 and 3.5. In general, the higher the content of heavy elements such as iron and lead, the higher the specific gravity of the mineral. For example, the specific gravity of magnetite ($Fe_3O_4$) is 5.2; that of galena (PbS) is 7.5. Gold, at 19, has the highest specific gravity of any mineral.
7. Other physical properties are sometimes useful in mineral identification (table 1.3). For example, calcite is the only important mineral that dissolves in cold, dilute hydrochloric acid. Magnetite is the only mineral attracted to a small hand magnet, while halite has a unique salty taste. Micas are elastic when bent.

Minerals occur most commonly as aggregates composed of many crystals, rather than as single crystals. The most diagnostic properties of the common minerals are listed in table 1.4.

## *Economic Mineralogy*

A group of about 20 minerals includes probably 99%, by volume, of those present in the Earth's crust. These abundant minerals predominate in the common rocks. But more than 3,500 different minerals are known, and many of the less common ones are important sources of chemical elements needed in our industrial civilization (table 1.5). Others are used for decorative purposes, such as gems in jewelry. Many cities have gem and mineral societies that organize meetings and exhibitions to display and trade semiprecious gemstones.

## *Environmental Aspects of Minerals*

Some minerals can cause environmental problems because of their physical properties and chemical compositions—for example, calcite, halite and gypsum, pyrite, and clay minerals.

Calcite is the essential mineral in *limestone* and is also one of the more easily dissolved of the abundant minerals. Because limestones are such widely distributed

TABLE 1.3

Mineral Classification Chart

| Metallic Luster | Other Characteristics | Mineral |
|---|---|---|
| **Gray streak** | Perfect cubic cleavage; H = 2.5; heavy, sp. gr. = 7.6; silver gray color | Galena<br>PbS |
| **Black streak** | Magnetic; black to dark gray; H = 6; sp. gr. = 5.2; commonly occurs in granular masses; single crystals are octahedral | Magnetite<br>$Fe_3O_4$ |
| **Gray to black streak** | Steel gray; soft, smudges fingers and marks paper, greasy feel; H = 1; sp. gr. = 2; luster may be dull | Graphite<br>C |
| **Greenish black streak** | Golden yellow color; may tarnish purple; H = 4; sp. gr. = 4.3 | Chalcopyrite<br>$CuFeS_2$ |
| | Brass yellow; cubic crystals; common in granular aggregates; H = 6–6.5; sp. gr. = 5; uneven fracture | Pyrite<br>$FeS_2$ |
| **Reddish brown streak** | Steel gray, black to dark brown, red to red-brown streak; granular, fibrous, or micaceous; single crystals are thick plates; H = 5–6; sp. gr. = 5; uneven fracture | Hematite<br>$Fe_2O_3$ |
| **Yellow-brown streak** | Yellow, brown, or black; hard structureless or radial fibrous masses; H = 5–5.5; sp. gr. = 3.5– 4 | Limonite<br>$FeOOH \cdot nH_2O$ |
| **Nonmetallic Luster—Dark Color** | | |
| **Harder than glass** — **Cleavage prominent** | Cleavage—2 directions nearly at 90°; dark green to black; short prismatic 8-sided crystals; H = 6; sp. gr. = 3.5 | Pyroxene Group<br>Complex Ca, Mg, Fe, Al silicates |
| | Cleavage—2 directions at approximately 60° and 120°; dark green to black or brown; long, prismatic 6-sided crystals; H = 6; sp. gr. = 3.35 | Amphibole Group<br>Complex Na, Ca, Mg, Fe, Al silicates |
| | White to gray; good cleavage in two directions at approximately 90°; striations on cleavage planes; H = 6; sp. gr. = 2.62–2.76 | Plagioclase Feldspar<br>$NaAlSi_3O_8$ to $CaAl_2Si_2O_8$ |
| **Harder than glass** — **Cleavage absent** | Various shades of green; sometimes yellowish; commonly occurs in aggregates of small glassy grains; transparent to transluscent; glassy luster; H = 6.5–7; sp. gr. = 3.5–4.5 | Olivine<br>$(Mg, Fe)_2SiO_4$ |
| | Red, brown, or yellow; glassy luster; conchoidal fracture resembles poor cleavage; commonly occurs in well-formed 12-sided crystals; H = 7–7.5; sp. gr. = 3.5–4.5 | Garnet Group<br>Fe, Mg, Ca, Al silicates |
| | Conchoidal fracture; H = 7; gray to gray-black; vitreous luster; sp. gr. = 2.65 | Quartz<br>$SiO_2$ |

TABLE 1.3—*Continued*

| | | | |
|---|---|---|---|
| Softer than glass | Cleavage prominent | Brown to black; 1 perfect cleavage; thin, flexible, and elastic when in thin sheets; H = 2.5–3; sp. gr. = 3–3.5 | Biotite<br>Hydrous K, Fe, Mg, Al silicate |
| | | Green to very dark green; 1 cleavage direction; commonly occurs in foliated or scaly masses; nonelastic plates; H = 2–2.5.; sp. gr. = 2.5–3.5 | Chlorite<br>Hydrous Mg, Fe, Al, silicate |
| | | Yellowish brown; resinous luster; cleavage in 6 directions; yellowish brown or nearly white streak; H = 3.5–4; sp. gr. = 4 | Sphalerite<br>ZnS |
| | | Four perfect cleavage directions; green through deep purple; transparent to translucent; cubic crystals; H = 4; sp. gr. = 3 | Fluorite<br>$CaF_2$ |
| | Cleavage absent | Red, earthy appearance; red streak; H = 1.5; sp. gr. = 5.26 | Hematite<br>$Fe_2O_3$ (earthy variety) |
| | | Yellowish brown streak; yellowish brown to dark brown; commonly in compacted earthy masses; H = 1.5; sp. gr. = 3.6–4.0 | Limonite<br>$FeOOH \cdot nH_2O$ |
| **Nonmetallic Luster—Light Color** | | | |
| Harder than glass | Cleavage prominent | Good cleavage in 2 directions at approximately 90°; commonly flesh-colored to dark pink; pearly to vitreous luster; H = 6; sp. gr. = 2.6 | Orthoclase feldspar:<br>$KAlSi_3O_8$ |
| | | Good cleavage in 2 directions at approximately 90°; white to gray; striations on some cleavage planes H = 6; sp. gr. = 2.62–2.76 | Plagioclase feldspars:<br>$NaAlSi_3O_8$ to $CaAl_2Si_2O_8$ |
| | Cleavage absent | Conchoidal fracture; transparent to translucent; vitreous luster; 6-sided prismatic crystals terminated by 6-sided triangular faces in well-developed crystals; vitreous to waxy; colors range from milky white, rose pink, violet, to smoky gray; H = 7; sp. gr. = 2.65 | Quartz<br>$SiO_2$ (silica)<br>Varieties:<br>Milky; Smoky;<br>Rose; Amethyst |
| | | Conchoidal fracture; variable color; translucent to opaque; dull or clouded luster; colors range widely from white, gray, red, to black; H = 7; sp. gr. = 2.65 | Microcrystalline Quartz $SiO_2$<br>Varieties:<br>Agate; Flint; Chert;<br>Jasper; Opal (amorphous) |
| | | Perfect cubic cleavage; salty taste; colorless to white; soluble in water; H = 2–2.5; sp. gr. = 2 | Halite<br>NaCl |
| | | Perfect cleavage in 1 direction; poor in 2 others; white; transparent; nonelastic; H = 2; sp. gr. = 2.3<br>Varieties:<br>Selenite: colorless, transparent<br>Alabaster: aggregates of small crystals<br>Satin spar: fibrous, silky luster | Gypsum<br>$CaSO_4 \cdot 2H_2O$ |

TABLE 1.3—*Continued*

| | | | |
|---|---|---|---|
| | | Perfect cleavage in 3 directions at approximately 75°; effervesces in HCl; colorless, white, or pale yellow, rarely gray or blue; transparent to opaque; H = 3; sp. gr. = 2.7 | Calcite $CaCO_3$ (fine-grained crystalline aggregates form limestone and marble) |
| | **Cleavage prominent** | Three directions of cleavage as in calcite; effervesces in HCl only if powdered; color variable but commonly white or pink; rhomb-faced crystals; H = 3.5–4; sp. gr. = 2.8 | Dolomite $CaMg(CO_3)_2$ |
| **Softer than glass** | | Good cleavage in 4 directions; colorless, yellow, blue, green, or violet; transparent to translucent; cubic crystals; H = 4; sp. gr. = 3 | Fluorite $CaF_2$ |
| | | Perfect cleavage in 1 direction, producing thin, elastic sheets; transparent and colorless in thin sheets; H = 2–3; sp. gr. = 2.8 | Muscovite $KAl_3AlSi_3O_{10}(OH)_2$ |
| | | Green to white; soapy feel; pearly luster; foliated or compact masses; one direction of cleavage forms thin scales and shreds; H = 1; sp. gr. = 2.8 | Talc $Mg_3Si_4O_{10}(OH)_2$ |
| | **Cleavage absent** | White to red; earthy masses; crystals so small no cleavage visible; becomes plastic when moistened; earthy odor; soft; H = 1.2; sp. gr. = 2.6 | Kaolinite $Al_2Si_2O_5(OH)_4$ |

rocks, dissolution of the ground surface and shallow subsurface is a common phenomenon in many areas. Rainwater seeps into cracks in the limestone and within a few hundred to a few thousand years can create large holes in the rock. When this occurs at shallow depths—perhaps a few tens of feet below the ground surface—it produces caverns such as Carlsbad Cavern in New Mexico. The ground above such a cavern can collapse into it, carrying with it buildings, automobiles, and even people. Numerous cases of ground collapse have occurred in populated areas of Florida and other southeastern states that have widespread, abundant limestone.

Halite and gypsum are much more soluble than calcite, but also much less abundant, so they cause fewer cases of ground collapse. However, they pose a great danger to the water supply in many regions, because even small amounts of sodium or sulfate ions in water can produce a salty taste (sodium) or a diuretic effect (sulfate). These problems are so severe in some areas, such as west Texas, that the residents must purchase bottled water imported from other regions.

Pyrite causes environmental problems in some of the mining areas where it is particularly abundant. The mineral dissolves and combines with oxygen in the atmosphere to form hydrogen ions, a major cause of acidic stream water.

$$\underset{\text{pyrite}}{2\,FeS_2} + \underset{\text{oxygen}}{9\,O_2} + \underset{\text{water}}{H_2O} \rightarrow \underset{\text{hematite}}{Fe_2O_3} + \underset{\text{sulfate ion}}{4\,SO_4^{-2}} + \underset{\text{hydrogen ion}}{2H^+}$$

Such waters, called *acid mine drainage,* cause serious pollution problems in some portions of the western United States.

Clay minerals (kaolinite, montmorillonite, and illite) and, to a lesser extent, micas can trigger environmental problems because their crystal structure causes easy cleavage. The sheetlike structure of these minerals permits rainwater to penetrate between the sheets, causing the sheets to swell and slip. Houses and other structures built on clay-rich rocks such as *shales* can have their foundations cracked by swelling clays. Buildings constructed on slopes underlain by shales sometimes slide downhill when the shale becomes saturated with water, a common phenomenon in hilly areas such as those of southern California.

TABLE 1.4

Diagnostic Characteristics of the Common Minerals

| Mineral | Characteristics |
|---|---|
| Quartz | Transluscent; conchoidal fracture; hardness 7 |
| Orthoclase feldspar | Right-angle cleavage; commonly pink; hardness 6 |
| Plagioclase feldspar | Right-angle cleavage; commonly white, striated |
| Biotite mica | Sheet structure and cleavage; dark green-black; hardness 2.5 |
| Muscovite mica | Sheet structure and cleavage; transluscent; hardness 2–2.5 |
| Hornblende | Two cleavages at 56° and 124°; elongate; dark green to black |
| Augite | Two cleavages at 87° and 93°; black |
| Olivine | Light green; conchoidal fracture |
| Chlorite | Sheet structure and cleavage; green; hardness 2–2.5 |
| Clay minerals | Microscopic crystals occurring as aggregates; hardness 2; kaolinite is white, montmorillonite and illite are green |
| Calcite | Dissolves with effervescence in dilute HCl; rhombohedral cleavage; normally white but other colors possible |
| Dolomite | Same as calcite but does not dissolve/effervesce unless powdered |
| Gypsum | Hardness 2; usually transluscent to white but other colors possible; three unequally good cleavages |
| Halite | Salty taste; three right-angle cleavages; hardness 2.5 |
| Hematite | Red-brown streak; metallic luster in visible crystals; earthy when microcrystalline |
| Garnet | Usually red but other colors possible; hardness 7; commonly has 12-sided crystal outlines |
| Kyanite | Bladed crystals; bluish; hardness 5 parallel to crystal length, 7 normal to length; vitreous-pearly luster |
| Sillimanite | Long slender crystals often in parallel groups; frequently fibrous; hardness 6–7 |
| Staurolite | Red-brown to black; resinous to vitreous luster; hardness 7–7.5; common interpenetrating right-angle twins |
| Epidote | Pistachio green; one perfect cleavage; hardness 6–7 |
| Magnetite | Highly magnetic; black; hardness 6 |
| Ilmenite | Like magnetite but nonmagnetic |
| Pyrite | Brassy yellow; typically in cubic crystals; streak greenish or brownish black |
| Graphite | Sheet structure and cleavage; readily soils fingers and marks paper (pencil "lead"); greasy feel; black |

TABLE 1.5

## Ore and Gem Minerals

| Element Recovered | Major Ore Mineral | Chemical Formula |
|---|---|---|
| Antimony | Stibnite | $Sb_2S_3$ |
| Arsenic | Orpiment, Realgar | $As_2S_3$, AsS |
| Beryllium | Beryl | $Be_3Al_2Si_6O_{18}$ |
| Chromium | Chromite | $FeCr_2O_4$ |
| Cobalt | Cobaltite | (Co, Fe)AsS |
| Copper | Chalcopyrite | $CuFeS_2$ |
| Iron | Hematite | $Fe_2O_3$ |
| Lead | Galena | PbS |
| Manganese | Pyrolusite | $MnO_2$ |
| Mercury | Cinnabar | HgS |
| Molybdenum | Molybdenite | $MoS_2$ |
| Nickel | Pentlandite | $(Fe, Ni)_9S_8$ |
| Tin | Cassiterite | $SnO_2$ |
| Titanium | Rutile | $TiO_2$ |
| Tungsten | Wolframite, Scheelite | $(Fe, Mn)WO_4$, $CaWO_4$ |
| Uranium | Carnotite | $K_2(UO_2)_2(VO_4)_2 \cdot 3H_2O$ |
| Vanadium | Carnotite | $K_2(UO_2)_2(VO_4)_2 \cdot 3H_2O$ |
| Zinc | Sphalerite | ZnS |
| Zirconium | Zircon | $ZrSiO_4$ |
| Native (uncombined) elements | | Gold, platinum, silver, sulfur |
| **Major Gem Minerals** | | |
| Aquamarine (blue beryl) | | $Be_3Al_2Si_6O_{18}$ |
| Chrysoberyl | | $BeAl_2O_4$ |
| Diamond | | C |
| Emerald (green beryl) | | $Be_3Al_2Si_6O_{18}$ |
| Garnet (common) | | $Fe_3Al_2(SiO_4)_3$ |
| Jade | | $NaAlSi_2O_6$ |
| Lapis lazuli | | $(Na, Ca)_4(AlSiO_4)_3(SO_4, S, Cl)$ |
| Olivine | | $(Mg, Fe)_2SiO_4$ |
| Opal | | $SiO_2 \cdot H_2O$ |
| Ruby (red corundum) | | $Al_2O_3$ |
| Sapphire (blue corundum) | | $Al_2O_3$ |
| Topaz | | $Al_2SiO_4F_2$ |
| Tourmaline | | Borosilicate of variable composition |
| Turquoise | | $CuAl_6(PO_4)_4(OH)_8 \cdot 4H_2O$ |
| Zircon | | $ZrSiO_4$ |

## Problems

1. Identify each of the minerals provided, specifying the particular physical properties you used in the identification. Be specific (e.g., green color rather than simply "color," two cleavages at right angles rather than only "cleavage").
2. The chemical formulas of solid compounds such as minerals are expressed by the smallest number of each type of atom necessary for there to be no net charge. By convention, the positively charged ion is listed first. For example, sodium chloride, the mineral halite, is written in chemical symbols as NaCl. Write the chemical formulas of each of the following minerals:
   a. sylvite, potassium chloride
   b. quartz, silicon dioxide
   c. corundum, aluminum oxide
   d. hematite, iron oxide (iron is +3)
3. Biotite mica and muscovite mica have identical crystal structures. Why, then, are they considered different minerals? Is this reason reflected in the criterion you used to distinguish between them? Why or why not?
4. The musical stage play *Oklahoma* contains a mention of a surrey with "isinglass windows you can roll right down." Isinglass is an obsolete name for one of the minerals in your tray. Identify it and state the physical property that enabled it to be used in lieu of glass, which was very expensive in the 1800s.
5. How might you distinguish between the smooth surfaces of a perfectly formed crystal and the cleavage surfaces of the same mineral?
6. Graphite is commonly used as a lubricant. Explain why this is possible, in terms of a visible physical property of the mineral.
7. What product in your house or dormitory might be made from each of the following minerals?
   a. graphite
   b. calcite
   c. garnet
   d. halite
   e. gypsum
   f. quartz
8. Most caves are formed in limestone, a rock composed entirely of the mineral calcite.
   a. Give two reasons why this is true.
   b. Quartz and feldspar are the two most abundant minerals exposed at and near the Earth's surface. Yet there are no commercial caves formed in rocks composed of these minerals. Why not?

Plagioclase showing typical white color and striations.

Orthoclase (microcline) showing typical pink color.

Quartz crystal with striations.

Hornblende crystals. Hornblende usually occurs as sand-size crystals in rocks and is identified by its cleavage.

Muscovite mica showing excellent sheetlike cleavage and translucency. Biotite mica is dark green.

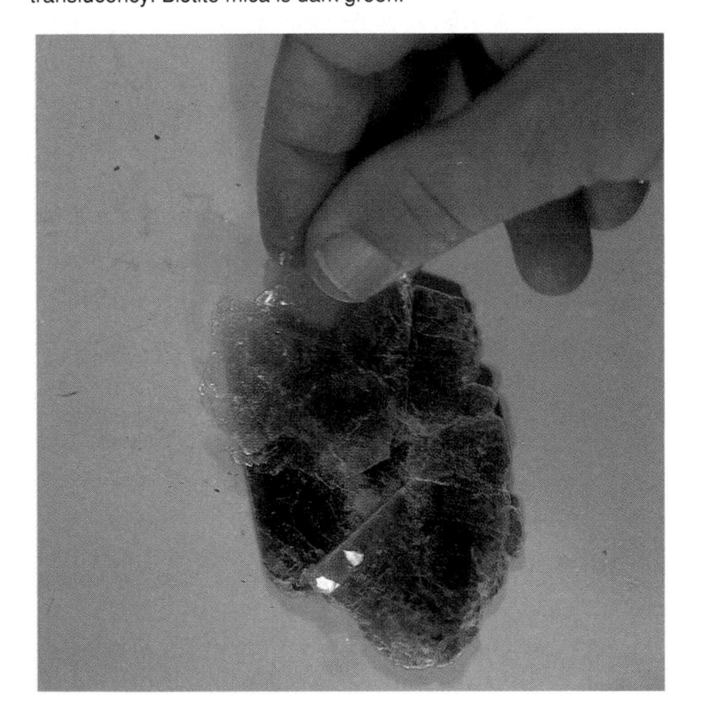

Augite, which usually occurs as sand-size crystals, is identified by its cleavage.

Garnet crystals and white plagioclase feldspar.

Gypsum (hardness of 2) is easily scratched by a fingernail (hardness of 2 1/2).

Calcite. Note double refraction and rhombohedral cleavage.

Cubic and rectangular cleavage fragments of halite.

Pyrite showing brassy yellow color.

## Further Reading/References


Bariand, P., Poirot, J-P., and Duchamp, M., 1992, *The Larousse Encyclopedia of Precious Gems,* New York, Van Nostrand Reinhold, 247 pp.

Chesterman, C. W., 1978, *The Audubon Society Field Guide to North American Rocks and Minerals.* New York, Alfred A. Knopf, 850 pp.

Gill, R., 1989, *Chemical Fundamentals of Geology.* Winchester, Massachusetts, Unwin Hyman, 291 pp.

Hurlbut, C. S., Jr. and Kammerling, R. C., 1991, *Gemology,* 2nd ed. New York, John Wiley & Sons, 336 pp.

Keller, P. C., 1990, *Gemstones and Their Origins.* New York, Van Nostrand Reinhold, 144 pp.

Klein, C., and Hurlbut, C. S., Jr., 1985, *Manual of Mineralogy,* 20th ed. New York, John Wiley & Sons, 596 pp.

Mitchell, R. S., 1979, *Mineral Names. What Do They Mean?* New York, Van Nostrand Reinhold, 229 pp.

Rice, P. C., 1987, *Amber: The Golden Gem of the Ages,* 2nd ed. New York, Kosciuszko Foundation, 289 pp.

Wilk, H., and Mederbach, O., 1985, *The Magic of Minerals.* New York, Springer-Verlag, 206 pp.


Galena.

Fluorite. Two octahedral crystals and one cubic cleavage fragment.

# Worksheet for Minerals

| Sample | Hardness | Cleavage (number and angular relationship) | Color | Streak | Luster | Other Distinctive Properties | Name of Mineral |
|---|---|---|---|---|---|---|---|
| | | | | | | | |
| | | | | | | | |
| | | | | | | | |
| | | | | | | | |
| | | | | | | | |
| | | | | | | | |
| | | | | | | | |
| | | | | | | | |
| | | | | | | | |
| | | | | | | | |
| | | | | | | | |
| | | | | | | | |
| | | | | | | | |
| | | | | | | | |
| | | | | | | | |
| | | | | | | | |
| | | | | | | | |

exercise

# TWO

## Igneous Rocks

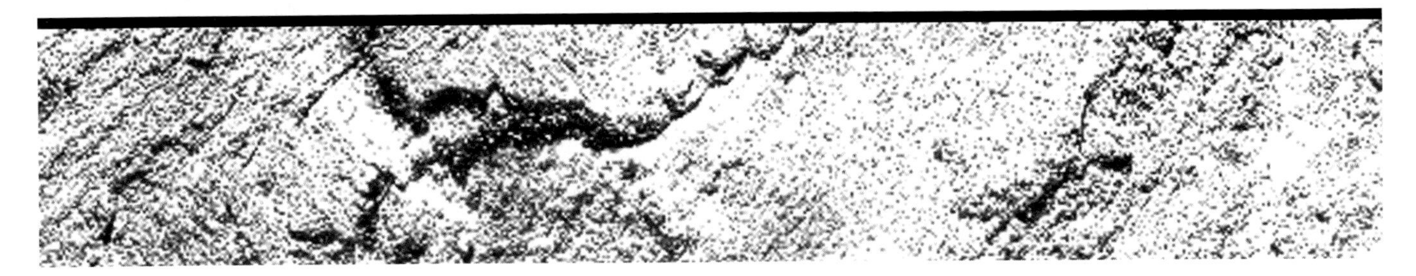

Geologists group rocks into three basic categories—*igneous, metamorphic,* and *sedimentary*—based on the processes by which they form. All three rock types grade into each other in appearance, but most rocks can be assigned easily to one type. Because of the different processes and temperature (and pressure) ranges in which they form, the types have distinctive physical appearances and often distinctive mineral compositions.

Igneous rocks are formed by solidification from molten material, nearly always at temperatures between 1200°C and 700°C, either beneath or above the ground surface. Molten material beneath the surface is termed *magma* and contains dissolved $H_2O$ molecules and gases as well as molten rock. When the magma breaks through the Earth's crust and flows onto the surface, it is called *lava.*

Magma forms at depths of many tens to several hundreds of kilometers. If it also cools and hardens slowly (crystallizes) at great depth, it forms *massive* (structureless) rock composed of fairly large, interlocking crystals. Crystals large enough to be seen by the unaided eye or with the help of a 10× magnifying lens are termed *phaneritic* (figure 2.1). To be visible, the crystals must be more than approximately 0.02 mm (0.001 in. in diameter). Smaller crystals cannot be distinguished by the human eye and are termed *aphanitic.*

As magma flows upward toward the surface and crystallization continues, the crystals produced become smaller. Therefore, hand specimens of many igneous rocks show fairly continuous variation in crystal size of their abundant minerals, from several centimeters to less than a few tenths of a millimeter. Such a rock texture is termed *seriate,* in contrast to an *equigranular* texture, in which all the crystals of the major minerals are roughly the same size. When deciding whether the texture of a rock is equigranular or seriate, examine the size range of only one mineral, such as orthoclase feldspar, at a time. Less abundant (*accessory*) minerals are always smaller in size than abundant ones, so if you do not consider mineralogy, you will identify all igneous rocks as seriate.

As the magma rises toward the surface, the temperature contrast between the melt and the surrounding rock increases, so crystallization becomes more rapid. At faster crystallization rates the newly forming crystals lack the time to grow very large; therefore, crystal size of major minerals is roughly correlated with their depth of crystallization. Very coarse crystals might have formed at depths of perhaps 25 km, finer crystals at any shallower depth. The fastest rate of cooling occurs when magma spills out onto the Earth's surface as lava, sometimes cooling instantly on

Figure 2.1

Classification of igneous rocks based on crystal size and mineral composition.

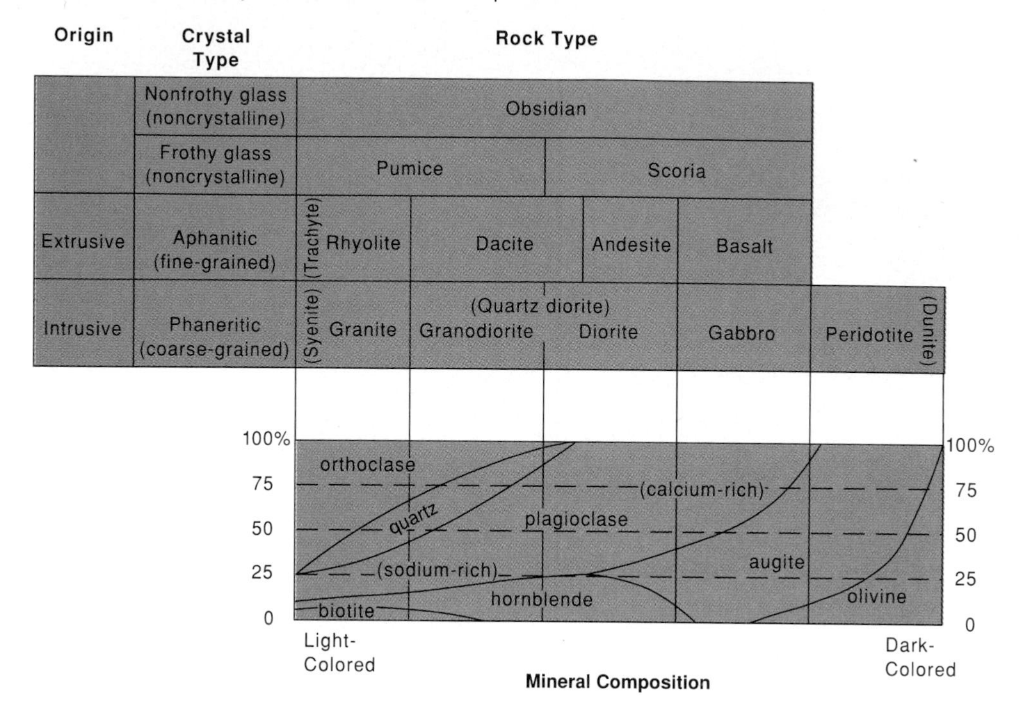


contact with seawater. Eruptions under the ocean and eruptions on Hawaii that flow into the ocean provide examples of such cooling. This extremely rapid cooling commonly freezes the atoms in the melt before they can associate into the regular, periodic structures called crystals. The resulting solid is noncrystalline (*amorphous*) and is termed *glass*. In hand specimens glass is distinguished from aphanitic rock by the appearance of its broken surface. Glass fractures produce a surface with concentric circles called *conchoidal fracture*, which looks similar to the fracture patterns of minerals that lack cleavage.

Although most magmas crystallize continuously, producing either an equigranular or a seriate texture (depending on whether the magma is moving upward as it crystallizes), some magmas crystallize at two distinctly different depths when they move rapidly from a deeper to a shallower location. The texture of the resulting rock exhibits two distinctly different crystal sizes and is termed *porphyritic*. The larger crystals are called *phenocrysts*; the smaller ones, the *groundmass*. As is evident from this description of crystallization patterns, equigranular texture grades into seriate, and seriate can grade into porphyritic. However, the igneous rocks used in the laboratory exercise are relatively clear examples of textural types.

Volcanic eruptions do not always occur as oozing flows of lava. Often magma contains sufficient gas to create an explosive eruption, as occurred at Mount St. Helens, Washington, in 1980 and at Mount Etna, Italy, in 1990. Such violent eruptions produce fragmental material called *tephra*; the layer of sediment produced is called *pyroclastic*, a term reflecting its origin in "fire." Pyroclastic sediments composed of fragments under 2 mm in size are called *tuffs*; if the fragments are coarser than 2 mm, the sediments are called *agglomerate*.

## *Mineral Composition*

Most igneous rocks contain at least 50% feldspar, with the remainder consisting of quartz, micas, hornblende, augite, and olivine. Certain minerals have a strong tendency to group together as magma crystallizes. One very common association is orthoclase feldspar with sodic plagioclase feldspar, quartz, and biotite mica. Another is plagioclase feldspar with augite and olivine. Always keep in mind, however, that nearly all things in nature are gradational, including mineral associations in rocks. Figure 2.1 illustrates the gradational character of the mineralogic changes seen in igneous rocks and the rock names used to describe them.

## *Igneous Rock Classification*

To classify an igneous rock you must identify both its texture and its constituent minerals.

*Step 1.* Is the rock glassy (vitreous luster, frothy appearance, conchoidal fracture) or crystalline?

*Step 2.* If the rock is crystalline, are the individual crystals visible to the unaided eye (phaneritic) or not (aphanitic)?

Granite. The pink mineral is orthoclase feldspar; the dark mineral is biotite mica; the clear grains are quartz; the white grains are plagioclase feldspar.

Diorite. White plagioclase feldspar with black hornblende.

Gabbro. The dark mineral is augite; the light mineral is calcium-rich plagioclase.

Peridotite. The dark mineral is augite; the green mineral is olivine.

Rhyolite. The crystals are too small to be visible to the unaided eye, but the pale pinkish color suggests the dominance of pink feldspar and quartz.

Rhyolite porphyry. White orthoclase feldspar crystals in brown aphanitic groundmass.

*Step 3.* If the rock is aphanitic, is it light-colored (high silica content) or dark-colored (high ferromagnesian-mineral content)?

*Step 4.* If the rock is phaneritic, are the abundant minerals (quartz, feldspars) about the same size (equigranular), gradational in size from small to large (seriate), or of two distinctly different sizes (porphyritic)?

*Step 5.* Identify the major minerals in the phaneritic rock and estimate their percentages. Identify the accessory (markedly less abundant) minerals and estimate their percentages.

*Step 6.* Place the rock in the appropriate pigeonhole and give it a name—for example, hornblende granodiorite, biotite granite, rhyolite porphyry, or diorite.

Basalt. The crystals are too small to be visible to the unaided eye, but the dark color suggests a large percentage of ferromagnesian minerals.

Obsidian glass, used by primitive peoples for making cutting tools and weapons.

Pumice, showing characteristic frothy vesicular texture. This glassy rock is used as an abrasive in Lava soap.

Scoria. Dark, vesicular rock formed as a glassy crust on basaltic lava flows.

## Problems

1. Identify each of the igneous rocks provided by determining its texture and mineral composition. Estimate the percentage of each mineral to see whether it agrees with the range shown in the igneous rock classification chart.
2. What can you infer about Earth history from the fact that a coarse-grained granite is exposed at the surface of the Earth?
3. Granite is the most common igneous rock in the upper 50 km of the Earth's crust. What can you infer about the relative abundances of the elements in this upper 50 km?
4. What inferences might you draw from the fact that igneous rocks show consistent mineral associations? For example, abundant quartz occurs with abundant orthoclase, and augite occurs with calcium-rich plagioclase.

## Further Reading/References

Barker, D. S., 1983, *Igneous Rocks.* Englewood Cliffs, New Jersey, Prentice-Hall, 417 pp.

Fisher, R. V., and Schminke, H.-U., 1984, *Pyroclastic Rocks.* New York, Springer-Verlag, 472 pp.

Mackenzie, W. S., Donaldson, C. H., and Guilford, C., 1982, *Atlas of Igneous Rocks and Their Textures.* New York, John Wiley & Sons, 148 pp.

Mitchell, R. S., 1985, *Dictionary of Rocks.* New York, Van Nostrand Reinhold, 228 pp.

Thorpe, R., and Brown, G., 1985, *The Field Description of Igneous Rocks.* New York, John Wiley, 154 pp.

## Igneous Rocks

| Sample Number | Texture | Percent Quartz | Percent Orthoclase | Percent Plagioclase | Type and Percent Accessories | Name of Rock |
|---|---|---|---|---|---|---|
| | | | | | | |
| | | | | | | |
| | | | | | | |
| | | | | | | |
| | | | | | | |
| | | | | | | |
| | | | | | | |
| | | | | | | |
| | | | | | | |
| | | | | | | |
| | | | | | | |
| | | | | | | |
| | | | | | | |
| | | | | | | |
| | | | | | | |
| | | | | | | |
| | | | | | | |
| | | | | | | |
| | | | | | | |
| | | | | | | |
| | | | | | | |
| | | | | | | |
| | | | | | | |
| | | | | | | |
| | | | | | | |

exercise

# THREE

## Sedimentary Rocks

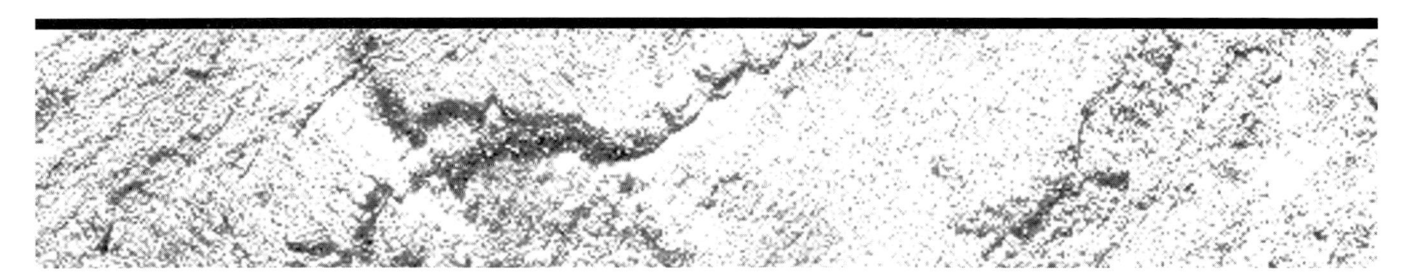

Sedimentary rocks are formed at the Earth's surface, excluding volcanic rocks. Following deposition, sediments are buried to depths at which the temperature might be as high as 250°C, a temperature reached at depths as shallow as 5 km or as deep as 25 km, depending on the *geothermal gradient* at the particular location. The geothermal gradient is the rate at which temperature increases with depth, which averages about 25°C/km. Some sedimentary rocks form by *cementation* of fragments of preexisting rocks; others form by precipitation of minerals from a supersaturated aqueous (water) solution. Sedimentary rocks are particularly important to environmental geologists because they underlie two-thirds of the Earth's land surface and are the host rocks for most of the economically important minerals (carnotite, gypsum, sulfur, copper, and others) and fluids (petroleum, natural gas, and water) on which our lives and modern civilization depend.

The most fundamental division of sedimentary rocks is into *clastic* (or *detrital*) rocks and *nonclastic* rocks. Clastic rocks are composed of fragments that have been buried and compacted, then glued together by the precipitation of mineral matter (*cement*) in the empty spaces (*pores*) among the fragments. Nonclastic rocks form entirely by precipitation from aqueous solution. Nonclastic rocks have a texture of interlocking crystals, like granite or basalt. Nonclastic rocks tend to be essentially monomineralic; *limestone*, for example, is composed largely or entirely of the mineral calcite.

### Conglomerates and Sandstones

About 85% of the sedimentary rocks found on the land surface are clastic rocks. In classification schemes we subdivide them first by fragment (grain) size, following an arbitrary scale developed in 1922 by C. K. Wentworth (table 3.1). Wentworth's scale was subsequently expressed in numbers called phi units ($\phi$) to facilitate statistical manipulations of grain-size data. If the fragments are mostly of gravel size (coarser than 2 mm or $-1\phi$), the rock is a *conglomerate*; if they are finer than 2 mm but coarser than 0.06 mm ($4\phi$), the rock is a *sandstone*; if they are finer than 0.06 mm, the rock is a *mudrock*. Less than 1% of all sedimentary rocks are conglomerates, about 20% are sandstones, and about 65% are mudrocks. Many clastic rocks contain mixtures of these three grain sizes and are called conglomeratic sandstone (gravelly sand), sandy conglomerate (sandy gravel), or muddy sandstone (muddy sand).

TABLE 3.1

Grain-Size Scale for Clastic Sediment and Sedimentary Rocks

| Name of Rock | Name of Sediment | Name of Grain | Grain Size (mm) | Grain Size (µm) | $\phi$ |
|---|---|---|---|---|---|
| Conglomerate | Gravel | Boulder | 256–4,096 | | −12 to −8 |
| | | Cobble | 64–256 | | −8 to −6 |
| | | Pebble | 4–64 | | −6 to −2 |
| | | Granule | 2–4 | | −2 to −1 |
| Sandstone | Sand | Very coarse sand | 1–2 | | −1 to 0 |
| | | Coarse sand | 0.5–1 | | 0 to 1 |
| | | Medium sand | 0.25–0.5 | 250–500 | 1 to 2 |
| | | Fine sand | 0.125–0.25 | 125–250 | 2 to 3 |
| | | Very fine sand | 0.062–0.125 | 62–125 | 3 to 4 |
| Mudrock | Mud | Coarse silt | 0.031–0.062 | 31–62 | 4 to 5 |
| | | Medium silt | 0.016–0.031 | 16–31 | 5 to 6 |
| | | Fine silt | 0.008–0.016 | 8–16 | 6 to 7 |
| | | Very fine silt | 0.004–0.008 | 4–8 | 7 to 8 |
| | | Clay | < 0.004 | < 4 | > 8 |
| | | | ↓ | ↓ | ↓ |

In geologic and environmental studies it is very useful to have a method for expressing the degree of grain-size variation in a clastic rock. This variation is called *sorting*. To a geologist, a well-sorted rock is one in which the grains are all about the same size. To a civil engineer, however, a well-sorted rock is one with a *wide* variation in grain size. The reason for this difference is that geologists are concerned with the efficiency of the natural processes that work to narrow the range of sizes, whereas civil engineers assess the value of grain aggregates for use in manufacturing concrete, in which a wide range of grain sizes is desirable. Hence, good sorting to a geologist is poor sorting to a construction engineer. Degree of sorting is also very important in relation to groundwater flow; groundwater flows more easily through sandstones with a narrow range of grain sizes.

The sand- and gravel-size grains in sandstones and conglomerates are composed almost entirely of three types of particles: quartz, feldspars, and undisaggregated rock fragments (figures 3.1 A-D). A sandstone composed of more than 90% quartz is called a *quartz sandstone*. If quartz is less than 90% and feldspar is more abundant than undisaggregated rock fragments, the rock is called a *feldspathic sandstone* or *arkose*. If quartz is less than 90% and undisaggregated rock fragments are more abundant than feldspar, the rock is called a *lithic sandstone*. A lithic sandstone that contains at least 10% clay is called a *graywacke*. The analogous terms for conglomerates are quartz conglomerate, feldspathic or arkosic conglomerate, and lithic conglomerate.

## Figure 3.1a

Pebble conglomerate. Pebbles are mostly quartz and chert; sand matrix is mostly quartz. Cement is quartz.

(a)

Loose grains deposited by water and wind are eventually buried by newly deposited sediment to depths of up to many tens of kilometers. During burial the grains are forced closer together (*compacted*), and water flows through the sediment, moving through the pore spaces among the irregularly shaped grains. As temperature increases during burial and chemical conditions change, crystals of various minerals precipitate in the pore spaces, creating the intimate crystal-to-crystal contact that lithifies the sediment into a hard rock. The common cement precipitates are calcite, quartz, clay, and hematite.

The clastic grains that eventually become sedimentary rock are deposited by water or wind in layers. The layers are bounded by surfaces of discontinuity called *bedding* or *lamination* surfaces. For some environmental concerns, such as the availability of groundwater, not only the thickness of the layers is important, but also their continuity. Some layers extend only for a few meters, such as a sand bar in a river channel; others extend for many kilometers, such as a linear beach along the Atlantic coast of the United States. Continuity of layers is commonly controlled by the environment in which the sand is deposited. In ancient rocks layers are more likely to be tabular in plan view than linear. The difference between the linear trend of modern sediments and the more tabular aspect of ancient sediments

## Figure 3.1b–d

Coarse-grained sandstones of varying composition. (b) is all clastic quartz, cemented by quartz, so the rock is very light-colored; (c) is rich in feldspar (white grains) as well as quartz, and is cemented by hematite (reddish color); (d) contains abundant dark-colored rock fragments and is cemented by clay (grains too small to be visible). All three sandstones are fairly well sorted and may have some porosity. Most sand grains are about 1 mm in diameter; scale is in centimeters.

(b)

(c)

(d)

**Figure 3.1e**

Shale showing excellent fissility; it breaks easily along closely spaced parallel surfaces because of parallel alignment of its clay minerals. Dark color reflects the presence of several percent organic matter. Hand lens is about 2 cm in diameter.

(e)

**Figure 3.2**

Exposed surface of bentonite illustrating "popcorn topography." Length of hammer is 28 cm.

reflects the shifting geographic location of depositional environments. For example, global warming might be partially responsible for both melting of the continental glaciers in polar regions and thermal expansion of ocean water, resulting in a worldwide sea-level rise of 2.4 ± 0.9 mm/yr (approximately 1 in/10 yr). Although this rise might seem insignificant, it is a major factor behind severe coastal-erosion problems in low-lying areas along the east coast of the United States. This sea-level rise causes the shoreline to move landward, creating a tabular unit of beach sand like those seen in ancient rocks.

Although mean sea level is rising around the world, it is dropping in some particular areas. This drop might be caused either by rebound of the Earth's crust following removal of the weight of glacial ice (as in Scandinavia), which can overcome the sea-level rise caused by the melting; or by upward movement of the land surface, caused by processes in the upper mantle of the Earth, immediately below the crust (as in California). If the land rise exceeds the sea-level rise, then the linear unit of beach sand will expand seaward rather than move landward.

## *Mudrocks*

Clastic rocks composed of grains smaller than 62 μm (4ϕ) are called mudrocks (figure 3.1E). They can contain various proportions of silt- and clay-size debris, and the relative amounts of each size are important to environmental scientists. Most silt grains are composed of quartz; clay-size grains are primarily clay minerals. Because the crystal structure of clay minerals is like that of micas—sheetlike with cleavage parallel to the sheets—they have a strong tendency to slip past each other on slopes, facilitating large-scale sediment movement. Also, the clay mineral called *montmorillonite* swells a great deal on contact with water, causing much structural damage to buildings and foundations constructed without proper soil studies. When a bed composed entirely of montmorillonite dries out, it can form a clay-ball pattern known as *popcorn topography* (figure 3.2).

If montmorillonite is abundant in a soil, the soil swells or shrinks, depending on the availability of water, and with devastating results to buildings (figure 3.3). Such soils are termed *vertisols* by soil scientists and *expansive* or *swelling soils* by engineers. Measured free-swelling ranges are from 50% to 2,000%, and the expansive pressure generated ranges from 15,000 kg/m$^2$ to 50,000 kg/m$^2$. This pressure is much greater than the load imposed by small buildings, which can be uplifted and rotated by the expanding soil. Buried water mains and utility lines can be similarly affected.

Mudrocks in which the clay flakes are well aligned display *fissility* (figure 3.1E), a tendency to split along very closely spaced, parallel planes. Most mudrocks are fissile and are called *shale*. Those that are not fissile are called

## Figure 3.3a

Examples of soil problems. (a) Slab poured in dry season, soil expansion at periphery during wet season; (b) slab poured in wet season, soil shrinkage at periphery in dry season; (c) building supported by cut and fill subject to differential expansion and contraction.

*(a) From Gary B. Griggs and John A. Gilchrist,* Geologic Hazards, Resources, and Environmental Planning, *2d ed. Copyright © 1983 Wadsworth Publishing Company, Belmont, CA. Reprinted by permission.*

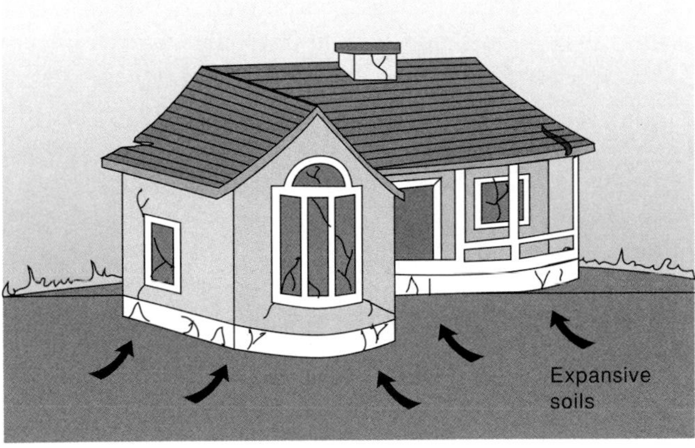


(a)

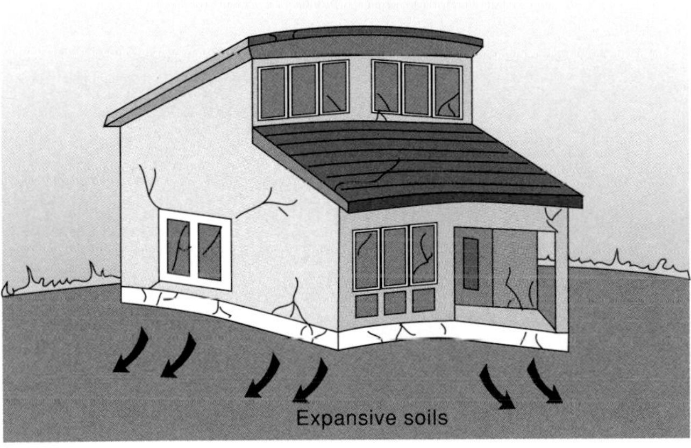


(b)

(c)

## Figure 3.3b

Building wall damaged by stair-step corner fracturing, Lakewood, Colorado, October 1976. The center section of the wall at right has been uplifted relative to the corner at left. Damage of this sort is frequently caused by expansive clay.

*mudstones*. Some investigators refer to all mudrocks as shales, using the term to denote all fine-grained clastic rocks rather than just mudrocks with fissility.

### *Limestones*

Rocks composed of either calcite ($CaCO_3$) or dolomite [$CaMg(CO_3)_2$] form about 15% of all sedimentary rocks. Those composed largely or entirely of calcite are called *limestones*; those made of dolomite are called *dolostones*. Dolostones form by chemical replacement of an original limestone and are less abundant than limestones. Many limestones contain some dolomite crystals, but most carbonate rocks are composed of either all calcite or all dolomite.

Limestones can be either clastic or nonclastic. Clastic limestones are the calcium carbonate equivalents of sandstones, with their sand- and gravel-size particles composed of calcium carbonate grains rather than silicate grains such as quartz and feldspar. The cement that holds the carbonate grains together is calcite (figure 3.4A). Nonclastic limestones are microcrystalline (aphanitic) (figure 3.4B); their individual crystals can be seen only through a microscope (figure 3.4B). In nearly all limestones both the coarse fragments and the microcrystalline calcite particles are deposited very near the place where they formed. This is in sharp contrast to sandstones, whose grains are usually transported great distances before being deposited.

## *Less Abundant Sedimentary Rocks*

Although they probably total no more than 1% of all sedimentary rocks, five other types are of sufficient geologic, economic, or environmental importance to warrant discussion.

*Evaporites* are deposits of very soluble minerals precipitated from evaporating pools of saline water. Many ancient evaporite deposits cover extensive areas but are visible as outcrops only in arid regions. In more humid climates they dissolve and are covered by soils. The most common evaporite minerals are halite (NaCl) and gypsum ($CaSO_4 \cdot 2H_2O$), both of which are of considerable economic importance. Other economically important evaporites are trona, a hydrated sodium carbonate mineral used in the manufacture of bicarbonate of soda; borax, a complex borate mineral used in household cleansing products (20 Mule Team Borax, Boraxo); and sylvite (KCl), used as a substitute for sodium chloride for people on "salt-free" diets.

*Cherts* are beds or nodules of microcrystalline quartz (figure 3.4C) formed by the recrystallization of the skeletons of microscopic, one-celled plants (diatoms) and animals (radiolaria). Chert is (micro-) equigranular and occurs in a variety of colors, including *jasper* (red) and *flint* (black). *Agate* is a variety of fibrous chert (variegated).

*Phosphorites* are beds composed of clastic or nonclastic particles of apatite. Phosphorites are dark-colored and typically contain both organic matter and commercially important amounts of uranium and vanadium. Phosphorites form largely by chemical replacement of calcium carbonate crystals on the ocean floor. Ancient phosphorites are mined extensively in western Wyoming and Idaho for use as fertilizer.

*Taconite* is the commercial name for banded iron ores found in Michigan, Minnesota, and several provinces of eastern Canada. The bands consist of alternating layers of red chert (jasper) and one or more iron-rich minerals such as hematite or magnetite. These rocks formed early in Earth history, when the atmosphere had a very different composition from that of today. A less abundant type of iron ore, called *ironstone,* is composed of sand-size balls (ooids) of hematite (figure 3.4D).

*Organic rocks* composed of either plant debris or semisolid hydrocarbons are of ever-increasing economic importance. They include *coal, oil shale,* and *tar sand.* Coal is a black rock formed by compression and chemical alteration of land-plant debris during its burial to depths of several thousand meters. Oil shale is a brown to black, fine-grained, clastic, organic-rich rock that yields oil when heated. The organic matter is the remains of microscopic plants (algae). Tar sand is a sand or sandstone that contains hydrocarbons so viscous they cannot be removed by conventional petroleum-recovery methods. It forms when the easily evaporated fraction (volatiles) of the petroleum is lost from a petroleum accumulation.

## *Classification of Sedimentary Rocks*

To classify a sedimentary rock, you must identify both texture and mineral composition, as was the case for igneous rocks.

*Step 1.* If the rock is phaneritic, is it clastic or nonclastic? Nonclastic rocks have an interlocking texture like that of an igneous rock such as granite. The grains in clastic rocks do not interlock.

*Step 2.* If the rock is clastic, is it composed of quartz, feldspar, and rock fragments or is it composed of calcite and dolomite?

*Step 3.* If the rock is clastic, estimate both the mineral composition and the main grain sizes. If it is nonclastic, determine whether the rock is limestone or dolostone.

*Step 4.* If the rock is aphantic but not limestone or dolomite, is it fissile or nonfissile?

*Step 5.* Name the rock (for example, medium-grained quartz sandstone, lithic pebble conglomerate, fossiliferous clastic limestone, or black shale).

## Figure 3.4a

Limestone consisting of fossil shells about 2 cm in length set in a matrix of calcite crystals about 0.5 mm in diameter.

## Figure 3.4b

Laminated microcrystalline limestone. Identified by its softness and the fact that if fizzes when acid is dropped on it.

## Figure 3.4c

Chert, an aphanitic rock composed of $SiO_2$. Although it looks like aphanitic limestone, chert is very hard and does not react in acid.

## Figure 3.4d

Rock formed of ooids of hematite ($Fe_2O_3$). The color is very distinctive, as is the reddish-brown streak.

## Problems

1. On the following chart, describe and name each of the sedimentary rocks in your tray.
2. Which do you think is likely to contain more pore space: a sandstone, a shale, or a limestone? Why?
3. Sand grains are usually deposited in a different geographic location than are clay minerals. Why do you think this is true?
4. Sedimentary rocks rarely contain both abundant quartz sand and abundant clastic limestone fragments. Can you explain this fact? (Hint: recall the physical properties of the two types of particles.)
5. Micas and clay minerals can be compacted more efficiently than can quartz and feldspar grains. Why?
6. Do you think fragments of chert in a sandstone could be compacted to the degree possible with shale fragments? Explain your answer.
7. Suppose a feldspar grain and a gypsum grain were being transported in a stream. Which grain do you think would survive longer? Why?
8. Would you rather build your house on a quartz-cemented sandstone or on a calcite-cemented sandstone? Why?
9. Would you rather build your house on a gentle slope underlain by shale or on a somewhat steeper slope underlain by a quartz-cemented quartz sandstone? Explain your answer.
10. Do you think halite would make a durable tombstone in a desert climate? Why or why not?
11. On which of the three major types of sedimentary rocks would you choose to locate a garbage dump (sanitary landfill)? Why?

## Further Reading/References


Blatt, H., Middleton, G. V., and Murray, R. C., 1980, *Origin of Sedimentary Rocks,* 2nd ed. New York, Prentice-Hall, 782 pp.

Dietrich, R. V., 1989, *Stones: Their Collection, Identification, and Uses,* 2nd ed. Prescott, Arizona, Geoscience Press, 191 pp.

Jochim, C. L., 1981, *Home Landscaping and Maintenance on Swelling Soil.* Colorado Geological Survey Special Publication 14, 31 pp.

Leeder, M. R., 1982, *Sedimentology, Process and Product.* Boston, George Allen & Unwin, 344 pp.

Pettijohn, F. J., 1975, *Sedimentary Rocks,* 3rd ed. New York, Harper & Row, 628 pp.

Pettijohn, F. J., Potter, P. E., and Siever, R., 1987, *Sand and Sandstone,* 2nd ed. New York, Springer-Verlag, 553 pp.

Siever, R., 1988, *Sand.* New York, Scientific American Books, 240 pp.

Tucker, M. E., and Wright, V. P., 1990, *Carbonate Sedimentology.* Boston, Blackwell, 482 pp.

## Sedimentary Rocks

| **Sample Number** | **Clastic or Nonclastic** | **Crystal or Grain Size** | **Mineral Composition** | **Cement if Clastic** | **Rock Name** |
|---|---|---|---|---|---|
| | | | | | |
| | | | | | |
| | | | | | |
| | | | | | |
| | | | | | |
| | | | | | |
| | | | | | |
| | | | | | |
| | | | | | |
| | | | | | |
| | | | | | |
| | | | | | |
| | | | | | |
| | | | | | |
| | | | | | |
| | | | | | |
| | | | | | |
| | | | | | |
| | | | | | |
| | | | | | |
| | | | | | |
| | | | | | |
| | | | | | |

exercise

# FOUR

## Metamorphic Rocks

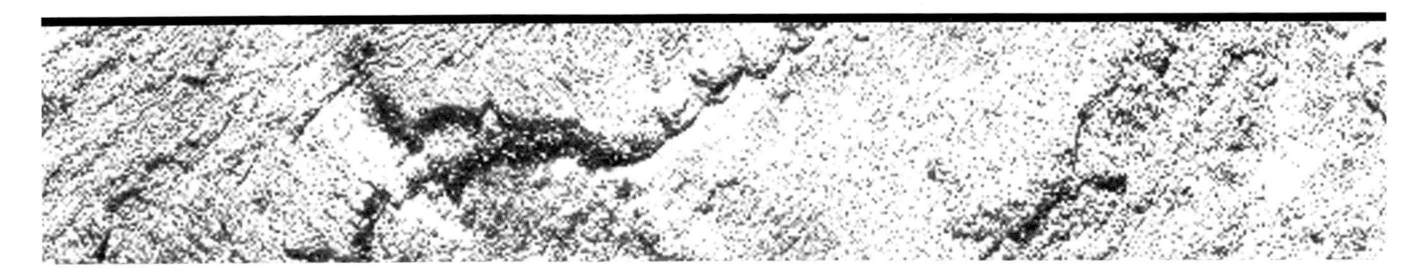

Most metamorphic rocks form at temperatures between about 250°C and 700°C, reached in the crust at depths of 10–50 km. These temperatures are too low to melt the rock and produce a magma, but high enough to add a great deal of energy to the chemical elements that make up the minerals. This energy increase permits the elements to move along the crystal surfaces in thin films of water and recombine into other minerals—minerals in equilibrium with the new temperature conditions. Many of these new minerals are distinctive to particular temperatures, but others are not. Kyanite and garnet, for example, are distinctive to metamorphic rocks, while quartz and feldspars form in both igneous and metamorphic rocks.

Pressure is a second factor that increases with depth and influences the formation of metamorphic rocks. Two types of pressure are differentiated: *hydrostatic* and *nonhydrostatic*. Hydrostatic pressure is equal in all directions, a condition characteristic of materials surrounded by a fluid. Minerals crystallize in a magma under hydrostatic pressure. Nonhydrostatic pressure, in contrast, varies in different directions. For example, when you place your hand on a deck of playing cards and move it parallel to the deck, a *shearing stress* results. The pressure is greater parallel to the deck than perpendicular to it, so the cards slide past each other. Metamorphic rocks can form under either hydrostatic or nonhydrostatic stress (or pressure), and the stress condition is reflected by the orientation of the minerals formed.

Minerals grown in metamorphic rocks that form under hydrostatic pressure have no preferred orientation, creating a texture like that of igneous rocks (table 4.1). Most metamorphic rocks, however, show a preferred orientation of minerals, indicating that they formed under nonhydrostatic stresses (table 4.2).

The most common preferred orientation is *foliation*, which results from a parallel arrangement of platy minerals, such as chlorite and micas, in the rock (figure 4.1). Three types of metamorphic rocks are foliated in this way: *slate*, *phyllite*, and *schist*. Slate looks much like shale except that its flat rock surfaces are more planar. Hand specimens of shale and slate have no noticeable differences except that slate surfaces are more intensely planar, a characteristic known as *slaty cleavage*. In many slate outcrops these planar surfaces do not lie parallel to the bedding of the rock unit. In both shales and slates, the crystals of clay or mica and quartz are too small to be visible (aphanitic) or to reflect light.

As temperature and nonhydrostatic stress increase during metamorphism, the platy minerals in slate grow to a large enough size to reflect light from their flat cleavage surfaces. At this point the rock becomes a *phyllite*. The crystals in a phyllite are still too small to be seen with the unaided eye, but their growth is made evident by the reflected light. Few new minerals are produced by this next grade of metamorphism; phyllites contain about the same minerals as slates, mostly quartz and muscovite mica. Phyllites form at between 200°C and 300°C.

TABLE 4.1

## Common Types of Nonfoliated Metamorphic Rocks

| Precursor Rock | Rock Name | Comments |
|---|---|---|
| Quartz sandstone | Quartzite | Composed of interlocking quartz grains |
| Conglomerate | Stretched-pebble conglomerate | Original pebbles distinguishable but strongly deformed |
| Basalt or gabbro | Greenstone | Composed of epidote and chlorite; green |
| | Amphibolite | Composed of amphibole and plagioclase; phaneritic |
| | Hornfels | Composed of pyroxene and plagioclase; aphanitic |
| Mudstone | Hornfels | Composed of quartz and plagioclase; aphanitic |
| Limestone/ dolostone | Marble | Composed of interlocking calcite or dolomite grains |
| | Skarn | Composed of calcite plus other minerals; multicolored |
| Peridotite or dunite | Serpentinite | Composed chiefly of serpentine; green |
| | Soapstone | Composed chiefly of talc; soapy feel |

TABLE 4.2

## Classification of the Common Foliated Metamorphic Rocks

| Increasing Grade of Metamorphism ↓ | Crystal Size | Rock Names | | Comments |
|---|---|---|---|---|
| | Microscopic, very fine-grained | Slate | | Slaty cleavage; well-developed planar surfaces |
| | Fine- to medium-grained | Phyllite | | Phyllitic texture well developed; silky luster |
| | Coarse-grained, macroscopic, mostly micaceous minerals; often with porphyroblasts | Schist | Muscovite schist<br>Chlorite schist<br>Biotite schist<br>Tourmaline schist<br>Garnet schist<br>Staurolite schist<br>Kyanite schist<br>Sillimanite schist<br>Hornblende schist | Types of schist named on the basis of mineral content |
| | Coarse-grained, mostly nonmicaceous minerals | Gneiss | | Well-developed color banding due to alternating layers of different minerals, most commonly quartz, feldspar, and ferromagnesian minerals |

## Figure 4.1a–g

a.–d. The important foliated metamorphic rocks. (a) Slate, identified by its aphanitic texture and very platy breakage pattern. (b) Phyllite, also aphanitic, but more micaceous, so a surface sheen of reflected light is present. (c) Schist, characterized by a predominance of coarse-grained micas. Dark red spots are garnet crystals. (d) Gneiss, composed of white feldspar, clear quartz, and dark streaks of biotite mica. Banding is diagnostic. (e) and (f) are equigranular quartzite and marble, respectively. They are the most common nonfoliated metamorphic rocks. (g) Natural asbestos, composed of three related fibrous minerals.

(a)

(b)

(c)

(d)

(e)

(f)

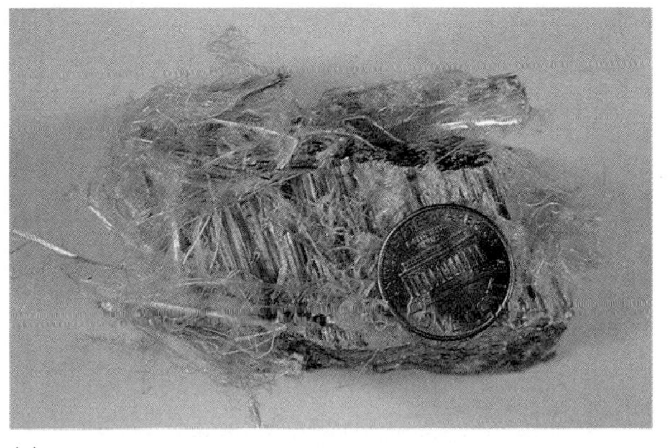

(g)

At temperatures above 300°C the micas grow large enough to be clearly visible to the unaided eye, and new minerals begin to form. The large, oriented micaceous minerals define the rock called *schist*. The first new sheetlike mineral to crystallize is green chlorite, which looks much like biotite mica. Additional muscovite and biotite then form, so that the most abundant minerals in the rock become chlorite, muscovite, and biotite. All of these minerals are sheetlike, so the foliation (also called *schistosity*) of the rock is very pronounced. The temperature of metamorphism at this point might be 350°C. From this temperature up to melting, the major change in the schist is the appearance of new minerals such as garnet, kyanite, staurolite, or sillimanite, each one marking the onset of a new range of temperature and pressure conditions.

In metamorphic rocks with micas less abundant than those in schists, orthoclase feldspar tends to be more abundant and the character of the foliation changes. Foliation in these rocks is represented by alternating bands of different minerals such as quartz, feldspar, and garnet. The rock is called a *gneiss*. Gneisses persist up to the melting temperature of the minerals in them, about 700°C, at which point they melt to produce a magma. Many magmas form from melting gneisses.

Nonfoliated metamorphic rocks are much less common than foliated ones. The abundant nonfoliated varieties include quartzite (composed largely or entirely of quartz), marble (composed largely or entirely of calcite or dolomite), and amphibolite (made up of hornblende and plagioclase feldspar). Occasionally these rocks are foliated.

### Metamorphic Rock Classification

Classifying a metamorphic rock requires recognizing its texture and mineral composition.

*Step 1.* Is the rock foliated or nonfoliated?

*Step 2.* If the rock is foliated, is it aphanitic, with slaty cleavage (slate)? Or aphanitic with a phyllitic sheen in reflected light (phyllite), visible micas in parallel orientation (schist), or foliation defined by bands of different mineral composition (gneiss)?

*Step 3.* If the rock is foliated and phaneritic, what are the abundant minerals? In schists the most common and abundant visible minerals are muscovite and biotite mica, quartz, epidote, garnet, kyanite, staurolite, and sillimanite. In gneisses the most common and abundant are quartz, feldspars, garnet, and micas.

*Step 4.* If the rock is nonfoliated, simply identify the one to three most important minerals.

*Step 5.* Name the rock. Typical names might be biotite-quartz schist, garnet-quartz-feldspar gneiss, or dolomite marble. Only the two or three most abundant minerals are normally used in the rock name, although all the minerals seen should be listed.

### Environmental Considerations

The most common environmental problems associated with metamorphic rocks are caused by foliation. Most metamorphic rocks contain abundant sheet-structure minerals such as chlorite and micas, and the cohesion between their adjacent grains is easily disrupted. A fairly small amount of water can cause slippage between mica flakes, so that mass movement of rock and overlying loose debris occurs on slopes. The movement is more rapid and noticeable on steep slopes but also occurs on gentle slopes, where it is more difficult to detect without sophisticated measuring techniques. Slippage problems are most intense when the hill slope lies parallel to the surface of foliation of the rock.

The implications of this loss of cohesion between adjacent mica grains are obvious. Buildings and other structures such as dams and highways built on micaceous metamorphic rocks, particularly on slopes, are inherently unstable. If their construction cannot be avoided, pilings should be sunk as deeply into the ground as possible to anchor the structures in rock below the zone of most intense weathering (water percolation).

## Problems

1. Identify each of the metamorphic rocks provided, by determining its mineral composition and whether it is foliated or nonfoliated.
2. High-grade gneisses and granites commonly have the same mineral composition. How can you distinguish between them?
3. How do you think the type of pressure or stress affects the variety of metamorphic rock formed?
4. Which type(s) of metamorphic rocks would be the best choices for roadbed material?
5. One type of foliated metamorphic rock has been widely used as a roofing material in private houses. Which one do you think it might be? The same rock was used in the last century to make writing tablets for schoolchildren.
6. Would a foliated or a nonfoliated metamorphic rock make the better tombstone? Explain your reasoning.

## Further Reading/References


Fry, N., 1984, *The Field Description of Metamorphic Rocks*. New York, John Wiley & Sons, 110 pp.

Gillen, C., 1982, *Metamorphic Geology*. London, George Allen & Unwin, 144 pp.

Mason, R., 1978, *Petrology of the Metamorphic Rocks*. London, George Allen & Unwin, 254 pp.

Yardley, B. W. D., 1989, *An Introduction to Metamorphic Petrology*. New York, John Wiley & Sons, 248 pp.

Yardley, B. W. D., Mackenzie, W. S., and Guilford, C., 1990, *Atlas of Metamorphic Rocks and Their Textures*. New York, John Wiley & Sons, 120 pp.

## Metamorphic Rocks

| Sample Number | Foliated? | Major Minerals | Name of Rock |
|---|---|---|---|
| | | | |
| | | | |
| | | | |
| | | | |
| | | | |
| | | | |
| | | | |
| | | | |
| | | | |
| | | | |
| | | | |
| | | | |
| | | | |
| | | | |
| | | | |
| | | | |
| | | | |
| | | | |
| | | | |
| | | | |
| | | | |
| | | | |
| | | | |

exercise

# FIVE

## Maps

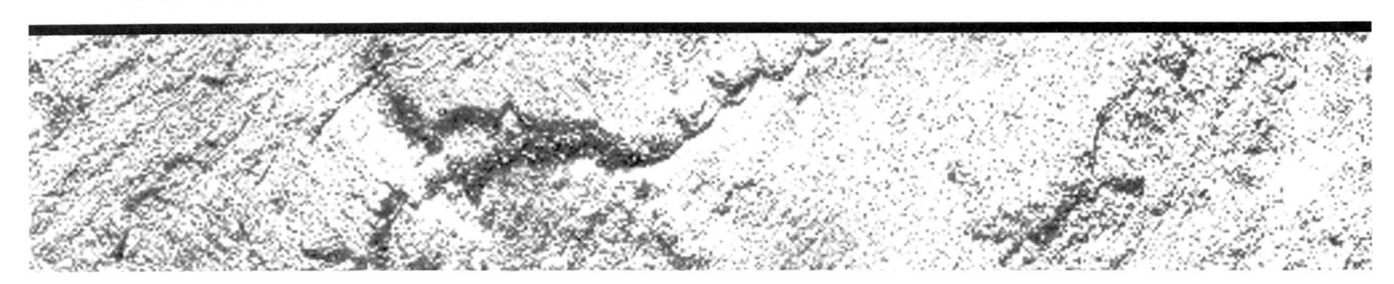

A map is a representation, usually on a flat surface, of a part of the Earth's surface. It can be an actual photograph taken from an orbiting satellite or an airplane, or a numerical approximation of the surface, with lines marking certain boundaries or elevations. Some maps show elevations above sea level (topographic maps) or below sea level (bathymetric maps); some show the contacts between rock units of geologic significance (geologic maps) or agricultural significance (soil maps). Still others are derivative maps constructed for environmental or engineering purposes—for example, flood-frequency maps, land-use maps, and maps of groundwater composition. In this and subsequent exercises we will consider several types of maps that are essential tools for environmental geologists.

### *Map Scale*

Features on a map are smaller than the actual features they represent. This reduction in size is termed the *scale* of the map. A scale of 1:100 means that one unit on the map is equal to 100 of the same units on the Earth's surface. In geological studies, a common scale for maps is 1:62,500, meaning that one inch on the map equals 62,500 inches on the ground—approximately one mile (1 mi = 63,360 in). A scale of 1:125,000 is a *smaller* scale than 1:62,500 because an inch represents two miles on the ground rather than one mile. A feature on the ground must be larger to be visible on this smaller-scale map. Many maps used in environmental studies have a numerical scale of 1:24,000 (one in. = 2,000 ft). In addition to a numerical scale, most maps have a graphic scale, a line or bar divided into segments that represent units of length on the ground.

### *Map Directions*

The Earth rotates around an imaginary line that passes through the Earth's center. The sites where this line intersects the surface of the Earth are called the *north* and *south geographic poles*, with geographic north shown on maps by an arrow. Also shown on geologic and many other types of maps is magnetic north, the direction to the *magnetic north pole*. Magnetic directions arise from the fact that the Earth acts like a simple bar magnet, with the imaginary bar passing through the center of the Earth and intersecting the surface near, but not at, the geographic poles. Many maps have arrows pointing to both the geographic and magnetic poles; the angle between them is the *magnetic declination*. When we use a magnetic compass to determine directions in the field, we adjust it to compensate for the magnetic declination, so that the needle points to "true" (geographic) north.

### *Map Coordinates*

Most maps useful for environmental studies are rather large-scale maps (they show a small area), making latitude and longitude inconvenient coordinates for identifying locations. In 1785, the federal government instituted the *Public Land Survey System* to deal with the problem of geographic location in land surveying during westward expansion (figure 5.1). The only states not subdivided according to the PLSS are Texas, because of its previous status as a separate nation, and the original thirteen colonies.

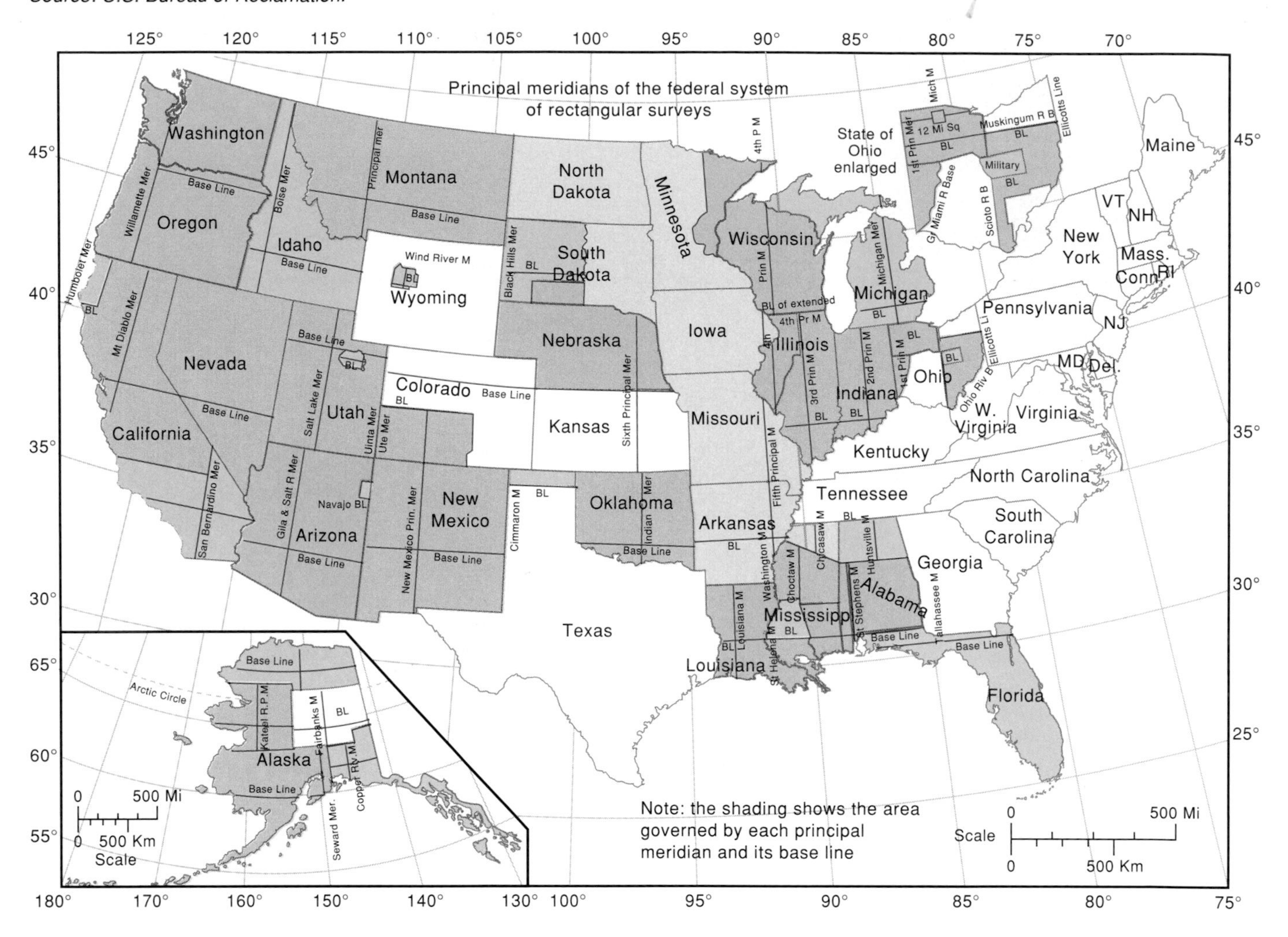


Figure 5.1

The Public Land Survey System.
*Source: U.S. Bureau of Reclamation.*

The PLSS was established in each state by surveying at least one east-west *base line* and one north-south *principal meridian* (figure 5.2). Adjacent states sometimes had the same base lines or principal meridians. Once the lines were established and related to latitude and longitude, additional lines parallel to them were drawn at six-mile intervals throughout each state, creating grids of squares, each square six miles on a side.

Squares along each east-west strip of the grid were called *tiers* or *townships* and were numbered, relative to the base line, as Township 1N, T2N, T3N (or T1S, T2S), etc. Squares along each north-south strip were called *ranges* and were numbered relative to the principal meridian (R1E, R2E, etc.).

Each of the six-mile squares (which, unfortunately, was also called a township) was further subdivided into 36 one-mile squares called *sections* and numbered in the rather peculiar pattern shown in figure 5.2B. In flat areas of the midcontinental United States, most of the main rural roads follow section lines. One section of land contains 640 acres. Subdivision of sections (figure 5.2C) was less formal and introduced no new terms. Reference might be made to the S 1/2 of a section, or perhaps to the SE 1/4 of a section. More specific locations were designated simply as quarters of quarter-sections—for example, the NW 1/4 of the NE 1/4 (an area of 40 acres)—and so on.

## *Elevations*

Elevations on maps are ultimately referenced to mean sea level as determined very accurately by the United States Coast and Geodetic Survey. More local reference points are *bench marks,* inland elevations determined accurately by the United States Geological Survey. Bench marks are widely scattered, so a particular map area might not contain

**Figure 5.2 a–c**

Generalized diagram of the PLSS. (a) Tier and range grid. (b) One township, 6 × 6 miles. (c) One section, 1×1 mile.

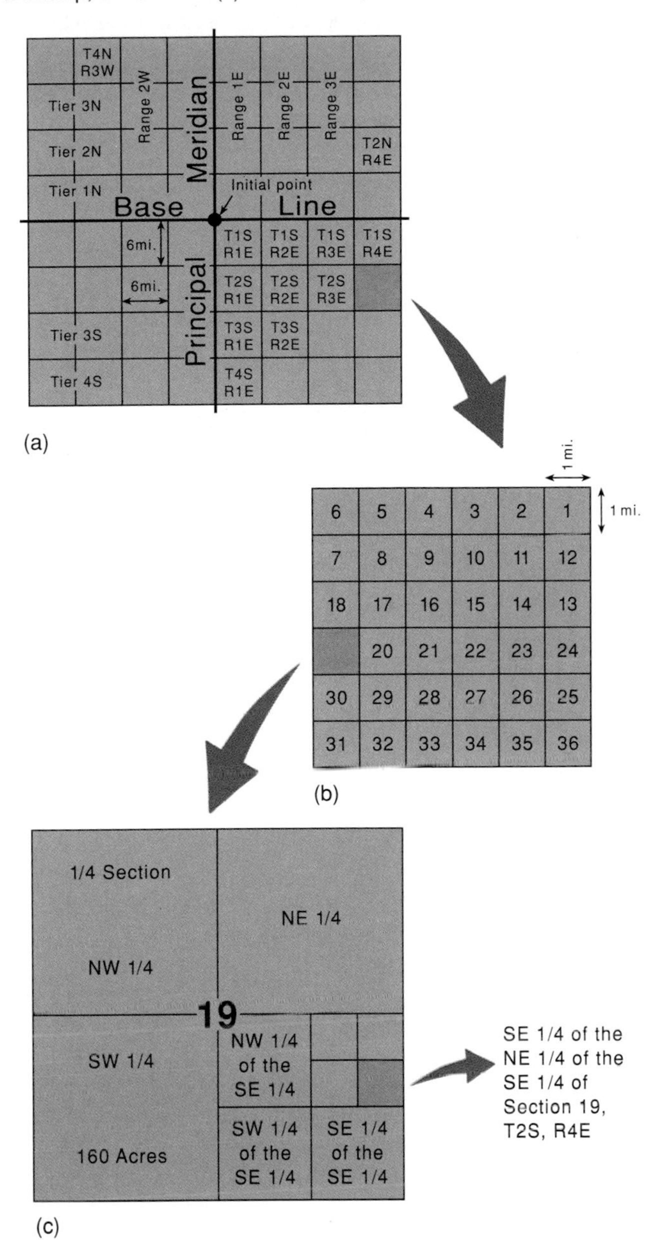


one. The map might, however, include points whose elevation above sea level has been fairly accurately determined.

## *Isopleths*

Isopleths are lines that connect points of equal value of a variable. Thus a topographic isopleth (or contour line) connects points of equal elevation above sea level; an *isocon,* points of equal salinity; an *isohyet,* points of equal precipitation; an *isopach,* points of equal thickness of a rock unit; an *isotherm,* points of equal temperature; and an *isoseismal line,* points of equal earthquake intensity. The difference in value between adjacent lines is called the *contour interval.* The choice of contour interval is arbitrary and is based on the scale of variation in the property being considered. For example, a map of an area in Kansas might require a topographic contour interval of 10 feet to show important features; in neighboring Colorado the contour interval might vary from 10 feet in the eastern part of the state to 100 feet in the western, more mountainous part.

## *Topographic Maps*

Topographic maps show the shape of the Earth's surface by means of contour lines. Variations in the surface shape help control streamflow rates, the likelihood of landslides, the probability of flooding, and many other things important to environmental geologists. A topographic contour line, which connects points of equal elevation above sea level, is formed by the intersection of an imaginary level surface with the ground (figures 5.3, 5.4). A natural example of a contour line is a shoreline around a lake. A man-made example of successive contours at different elevations is the pattern produced by contour plowing, a plowing pattern designed to reduce soil erosion.

Most precise locations do not fall exactly on a contour line—for example, a spring might be located between contour lines representing 430 and 440 feet above sea level. In such cases, we estimate the elevation by measuring with a graduated scale from the spring to one of the contour lines. If the spring is 65% of the distance from line 430 to line 440, for example, its elevation is approximately 436.5 feet. The elevation can only be approximated because this method assumes that the ground slopes evenly between the two contour lines. In reality, the surface might slope gently from 430 to 433 feet, then more steeply from 433 to 440 feet. In such a case, the method would yield a slightly erroneous, higher elevation for the spring. The accuracy of the estimate, therefore, depends on both the contour interval and the ground slope. A five-foot contour interval, rather than ten-foot, would reduce the potential for error.

*Relief* is defined as the difference in elevation between two points on a map. *Local relief* is the maximum difference in elevation within a designated small area on the map; *total relief* is the maximum difference in elevation between any two points on the map. Local relief helps control stream gradient and the frequency of large-scale earth movements such as landslides, and may also reflect relatively recent movement along breaks (faults) in the Earth's crust.

## *Topographic Profiles*

A topographic map provides a view from above the ground surface, using symbols and contour lines to show physical features and relief. This depiction permits us to visualize the shape of the ground surface and the location of buildings,

## Figure 5.3

**Rules for Contour Lines***

1. Every point on a *contour line* is the exact same elevation; that is, contour lines connect points of equal elevation.
2. Contour lines always separate points of higher elevation (uphill) from points of lower elevation (downhill). One must determine which direction on the map is higher and which is lower, relative to the contour line in question, by checking adjacent elevations.
3. The elevation between any two adjacent contour lines on a topographic map is the *contour interval*. Often every fifth contour line is heavier, so you can count by five-times the contour interval. These heavier contour lines are known as *index contours,* because they generally have elevations printed on them.
4. Contour lines never cross one another, except in one rare case: an overhanging cliff. In such a case, the hidden contours are dashed.
5. Contour lines merge to form a single contour line only where there is a vertical cliff.
6. Contour lines never split.
7. Contour lines that are
   a. evenly spaced indicate comparatively uniform slopes.
   b. closely spaced show steep slopes.
   c. widely spaced portray gentle slopes.
   d. irregularly spaced signify irregular slopes.
8. Contour lines form a V pattern when crossing streams. The apex of the V always points upstream (uphill).
9. A concentric series of closed contours represents a hill:
10. *Depression contours* have hachure marks on the downhill side, always close, and represent a closed depression:

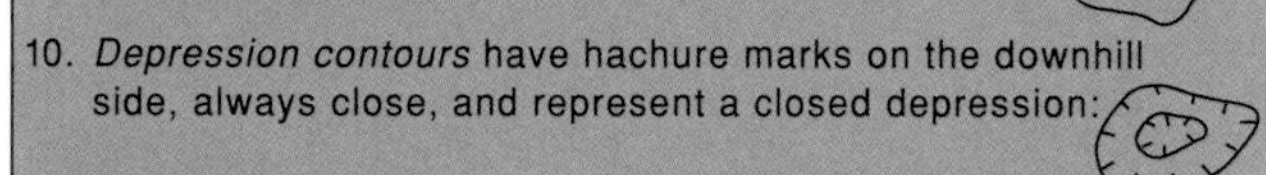

   a. Where the topography slopes downhill and a standard contour is adjacent to a hachured contour, the hachured contour is one contour interval lower than the standard contour.
   b. Where the topography slopes uphill and a standard contour is adjacent to a hachured contour, the two contours have the same elevation.
   c. Where one closed hachured contour encloses another closed hachured contour, the inner contour is one contour interval lower.
   d. Where a closed standard contour is enclosed by a hachured contour, they both have the same elevation.

*These same rules apply to bathymetric maps, which show the topography of the floors of lakes or oceans.

rivers, and other features. A *topographic profile* is a cross-section that shows the changing elevation of the land surface along a line. We select the position of the line either to provide a representative sample of the surface irregularities shown on the map or to show the configuration of the ground surface in an area of particular interest. Figure 5.5 illustrates the construction of a topographic profile (or cross-section). The procedure is as follows:

1. Draw the line along which you wish to construct the profile. Note the maximum topographic relief along the line.
2. Choose a vertical scale. Scales are arbitrary, but most have about 10 to 15 divisions between the lowest and highest elevations on the cross-section. Using fewer than 10 divisions risks losing desired details of the ground surface; using more than 15 adds more detail than is needed for most purposes. Choose a number of divisions that suits the intended purpose of the profile. Label the horizontal lines of the profile grid.
3. If the profile line runs exactly east-west, simply drop a dashed line from each point at which the profile intersects a contour line to the appropriate elevation on the profile grid, as shown in the illustration. Connect the dots with a smooth curve.
4. If the profile line does not run exactly east-west, place the upper edge of a clean strip of paper along the line. Where each contour line intersects

## Figure 5.4

The relationship between the shape of the land surface and contour lines on a topographic map. Steep slopes are shown by closely spaced contours, gentle slopes by widely spaced ones. In this example, the steep slope results from resistant layered rocks and the gentler slope from less resistant rocks such as shales. Note that when contours cross streams, they *V* upstream. The sand spit at the bottom of the map has no elevation that reaches 20 feet above sea level, so that no contour lines are present on the spit.
*Source: U.S. Geological Survey.*

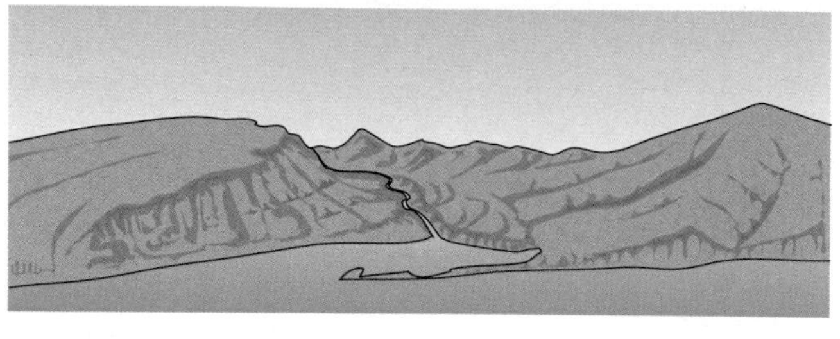

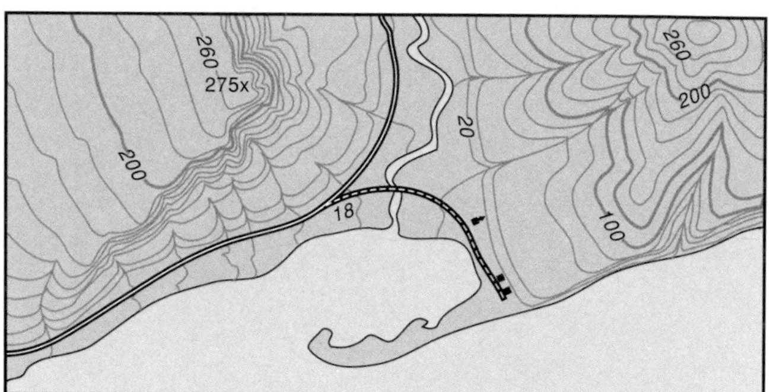

## Figure 5.5

Topographic profile across the Pseudomountain area along line A-A′. Vertical exaggeration = 32 ×.
*Top: Source: U.S. Geological Survey.*

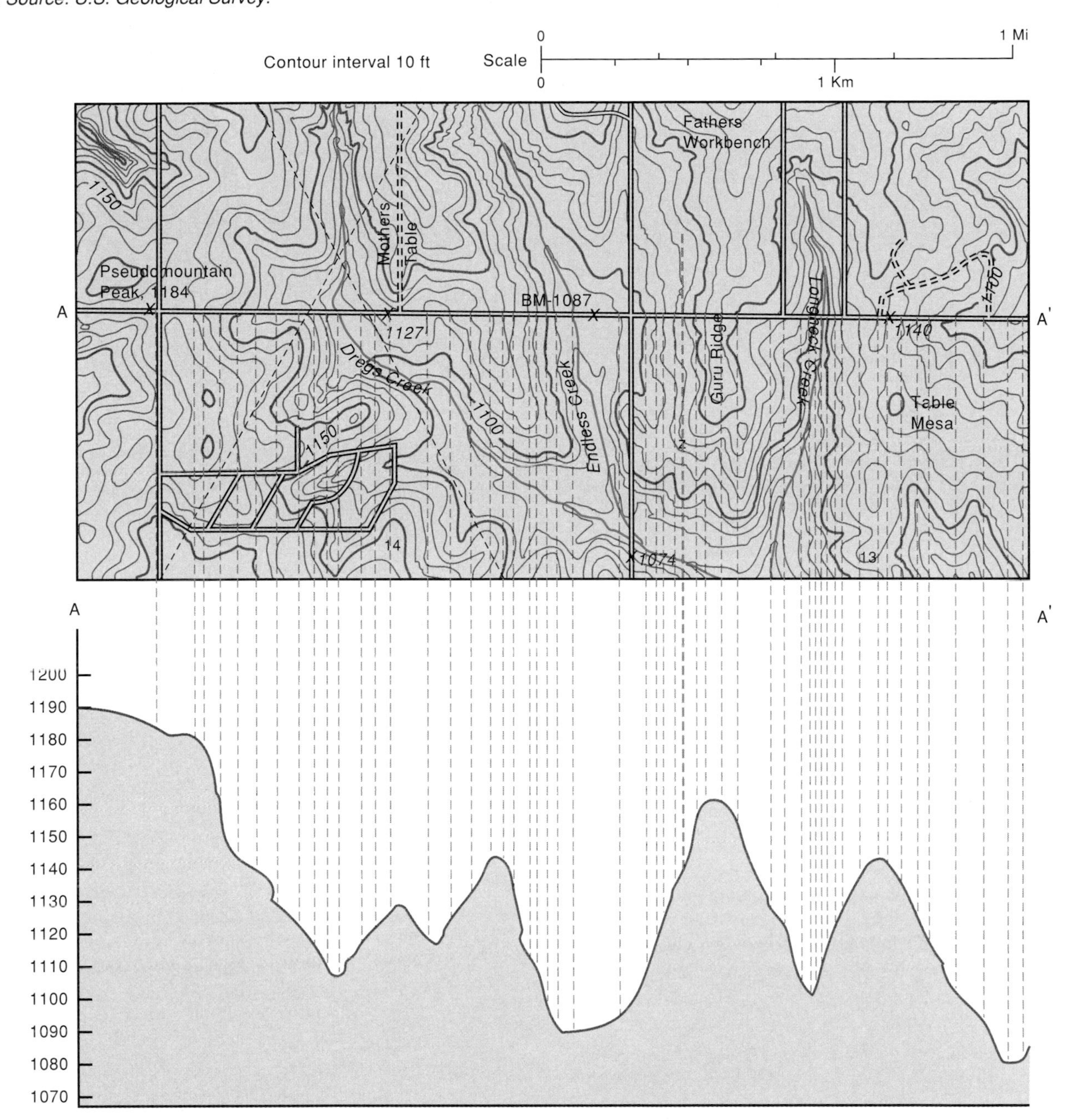


Vertical exaggeration = 32x.

the profile line, make a short tick mark on the paper and write the elevation by it. Then place the paper along the bottom edge of the profile grid and project from each tick mark up to the correct grid elevation.

A map is always smaller than the area it represents. Similarly, on any cross-section the vertical distance between two elevations on the paper is less than the true vertical distance (difference in elevation) on the ground. The vertical scale nearly always differs from the horizontal scale. The ratio between the fractional vertical scale and the fractional horizontal scale is called the vertical exaggeration. For example, if the horizontal scale is 1:62,500 and the vertical scale is 1:1,200, the vertical exaggeration is

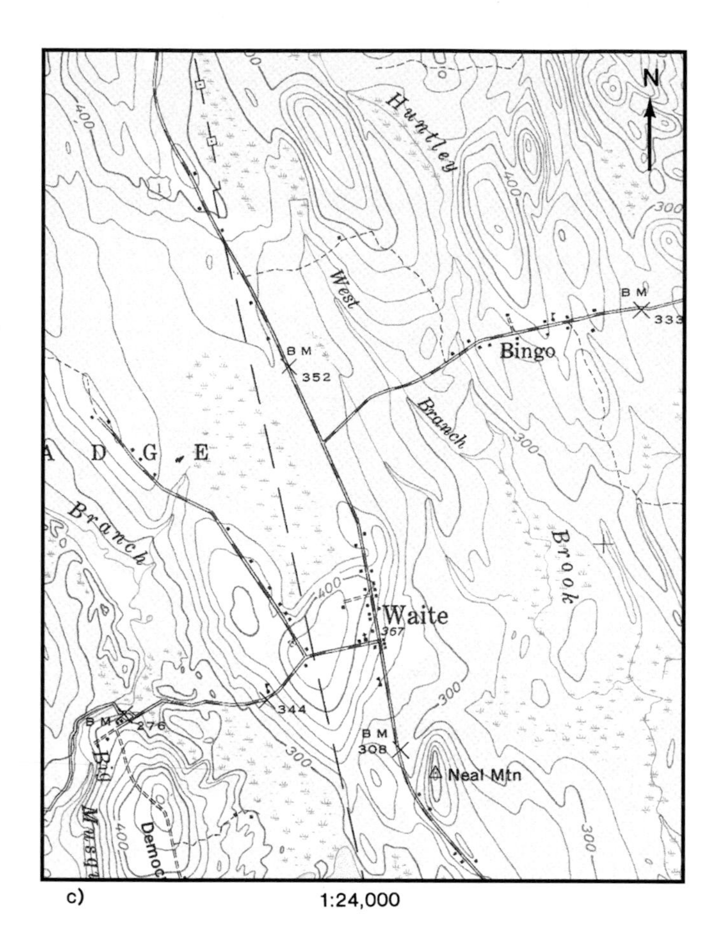

c) 1:24,000

approximately 52 × (62,500/1,200)—which means that the slopes on the topographic cross-section are 52 times steeper than the corresponding slopes on the ground.

## Problems

1. Make a topographic map by contouring the following points, using a 50-foot contour interval. Use only a pencil, and sketch lightly so you can erase errors easily. Numbers indicate elevation in feet above sea level for each adjacent dot. Use the contour provided on the map as a starting guide.

Shown above is a portion of the Waite 7 1/2′ topographic Quadrangle, Maine. Based on this map, answer the following questions.

2. How many square miles are shown on the map?
3. What is the straight-line distance from Waite to Bingo?
4. Why do you think the highway between the two towns was not built along the straight line?

Map for problem 1.

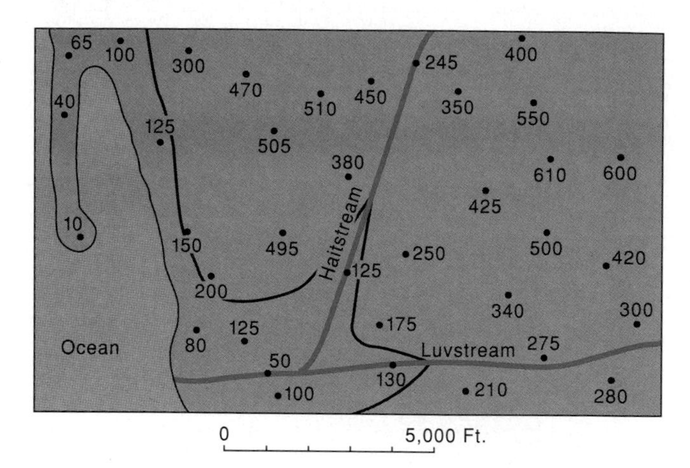


5. What is the total relief on the map?
6. In which direction does Huntley Brook flow?
7. What is the relationship between the direction in which streams intersect on the map and the direction in which they flow?
8. Construct a topographic cross-section from BM 276 to BM 333. What is the vertical exaggeration of your cross-section?

## Further Reading/References


Bart, H. A., 1991, A hands-on approach to understanding topographic maps and their construction: *Journal of Geological Education,* v. 39, pp. 303–305.

DeBruin, R., 1970, *100 Topographic Maps.* Northbrook, Illinois, Hubbard, 128 pp.

Drury, S. A., 1987, *Image Interpretation in Geology.* Winchester, Massachusetts, Allen & Unwin, 243 pp.

Miller, V., and Westerback, M. E., 1988, *Interpretation of Topographic Maps.* Columbus, Ohio, Merrill, 416 pp.

Muehrcke, P. and Muehrcke, J. O., 1992, *Map Use: Reading Analysis and Interpretation,* 3rd ed. Madison, Wisconsin, JP Publications.

Sabins, F. F., Jr., 1986, *Remote Sensing,* 2nd ed. New York, W. H. Freeman, 449 pp.

Thompson, M. M., 1981, *Maps for America,* 2nd ed. Reston, Virgina, U.S. Geological Survey, 265 pp.

Upton, W. B., Jr., 1970, *Landforms and Topographic Maps.* New York, John Wiley & Sons, 134 pp.

exercise

# SIX

## Geologic Maps

Topographic maps show the shape of the ground surface. Geologic maps, in contrast, show the areal distribution and orientation of the three-dimensional bodies of rock that appear on the surface. These surface appearances are termed *outcrops*. The entire Earth is, of course, underlain by rocks, but they are not always evident. Many humid tropical or subtropical areas, such as central Africa, the Amazon region of South America, and the southeastern United States, have few outcrops because their soils are so thick—sometimes tens of feet. In the northeastern and north-central United States and Canada, outcrops are scarce because they are covered by Pleistocene glacial debris. In contrast, outcrops are often both extensive and continuous in arid or semiarid regions such as those of the Middle East and western Texas. Where outcrops are scarce, geologists making a map infer probable rock distributions from surrounding areas with better outcrops. Thus, geologic maps are not maps of actual outcrops; rather, they incorporate both actual outcrops and geologic inference. A location shown as sandstone on a geologic map might in fact be a wheat field; we would find the sandstone only by digging through the soil.

Geologic maps show numerous features relating to the origin and geologic history of an area, many of them relevant to environmental concerns. Environmental geologists, therefore, must know how to read and interpret geologic maps. Geologic maps might suggest the best sites for placing sanitary landfills, drilling water wells, finding springs, or finding construction-grade sand and gravel, as well as reveal areas most vulnerable to landslides or earthquakes, or many other things of environmental importance. Special-purpose maps derived from geologic maps emphasize specific geologic hazards for engineering or environmental analyses.

What features related to rock distributions do geologic maps show (figures 6.1, 6.2)?

1. *Formations*. Geologists consider a formation to be a mappable rock unit defined by rock type (e.g., sandstone, shale), geologic age, fossil content, or some other characteristic that is both easy to see and diagnostic. Formations are given names, such as the Wellington Formation or Redwall Limestone, and assigned colors and symbols to identify them on a map. The Wellington Formation, which is of Permian age, might be assigned the symbol "Pw" (Permian, Wellington). Redwall Limestone might be "Mr" (Mississippian, Redwall). Different formations are shown on a map by colors or line patterns.
2. *Folds*. A fold is a bend in a layered rock. An "upfold" is called an *anticline*; a "downfold" is a *syncline*. In an eroded anticline the oldest layers are in the center of the fold, while in an eroded syncline the youngest layers are in the center.

3. *Faults.* A fault is a break in the Earth's crust along which movement has occurred. The rocks on one side of a fault might move up, down, or sideways with respect to the other side, but the movement is always parallel to the fault surface. Paved areas might crack, but only in movies does the Earth pull apart, normal to the fault, and swallow whole towns.
4. *Contacts.* A contact is the boundary between two rock units or between a rock unit and a fault.
5. *Strike.* The strike of a tabular rock layer is the compass direction of the line formed by the intersection of the layer with an imaginary horizontal plane (figure 6.3).
6. *Dip.* The dip of a rock layer is the vertical angle formed by the intersection of the layer with an imaginary horizontal plane (figure 6.3). The dip angle must be measured perpendicular to the strike direction; when measuring in any other direction the angle determined will be less than the true dip.
7. *Unconformities.* An unconformity is an ancient erosional surface covered by later sediment. The present ground surface is an unconformity in the making. Unconformities are normally identified in the map legend and might not be evident on the map itself.
8. *Other features.* A geologic map might also include some of the features shown on topographic maps, such as topographic or bathymetric contours, streams, or cities.

## *Environmental Geology Applications*

Geologists use the location and geologic age of map features to unravel the geologic history of the map area. Similarly, numerous features on geologic maps are important in environmental geology.

1. The lithology of a formation helps determine its ability to transmit such fluids as water, petroleum, and natural gas. A sand deposit, for example, has spaces between the grains that can fill with fluids, while a chemical precipitate such as gypsum does not.
2. Shale and clay deposits are rich in clay minerals. The platy shape of these minerals causes them to serve as slip surfaces on which landslides begin.
3. Faults often permit the movement of fluid through rocks that might otherwise be impermeable. The same is true of *joints,* rock fractures along which no movement has occurred. Unconformity surfaces can also serve as fluid conduits.

**Figure 6.1**

Geologic symbols for rock types.
*Source: U.S. Geological Survey.*

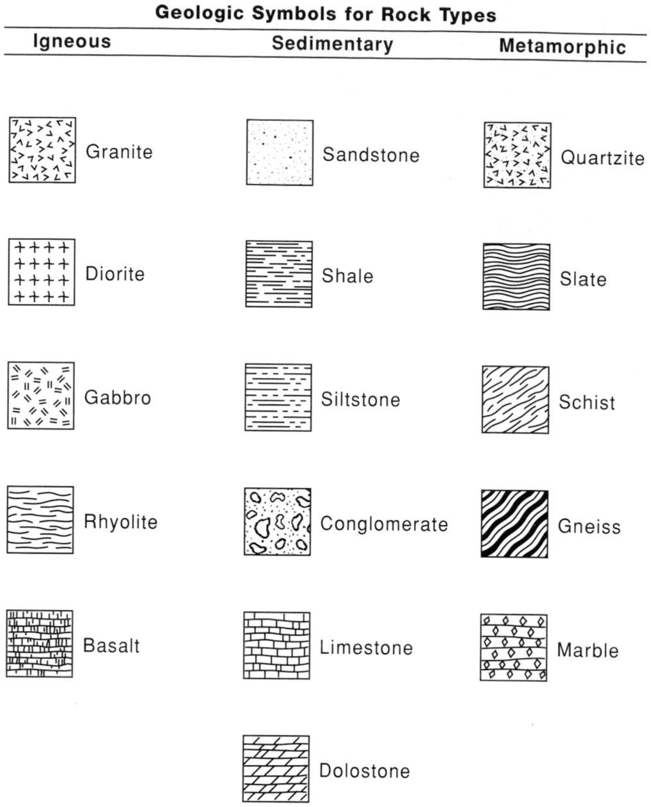

**Geologic Symbols for Rock Types**

| Igneous | Sedimentary | Metamorphic |
|---|---|---|
| Granite | Sandstone | Quartzite |
| Diorite | Shale | Slate |
| Gabbro | Siltstone | Schist |
| Rhyolite | Conglomerate | Gneiss |
| Basalt | Limestone | Marble |
| | Dolostone | |

4. The angle of dip of a fluid-containing layer helps determine the rate at which the fluid flows through the layer. Water flowing quickly through a rock can mean a more readily available supply for a nearby city.

Geologic maps are an essential tool for the environmental geologist. When combined with topographic maps and an understanding of rocks and minerals, they form the basis for making important decisions on environmental issues.

## *Geologic Cross-Sections*

A geologic map shows the locations of all the major rock units present on the ground surface, or that would be present if soil cover, glacial debris, and man-made structures were removed. But what happens below the ground surface? Does a water-bearing sandstone gradually become finer grained, grade into shale, and lose its water-bearing properties? What is the source of water leaking from a fault surface as a spring? Do limestones exposed at the surface

**Figure 6.2**

Symbols on geological maps published by the United States Geological Survey.
*Source: U.S. Geological Survey.*

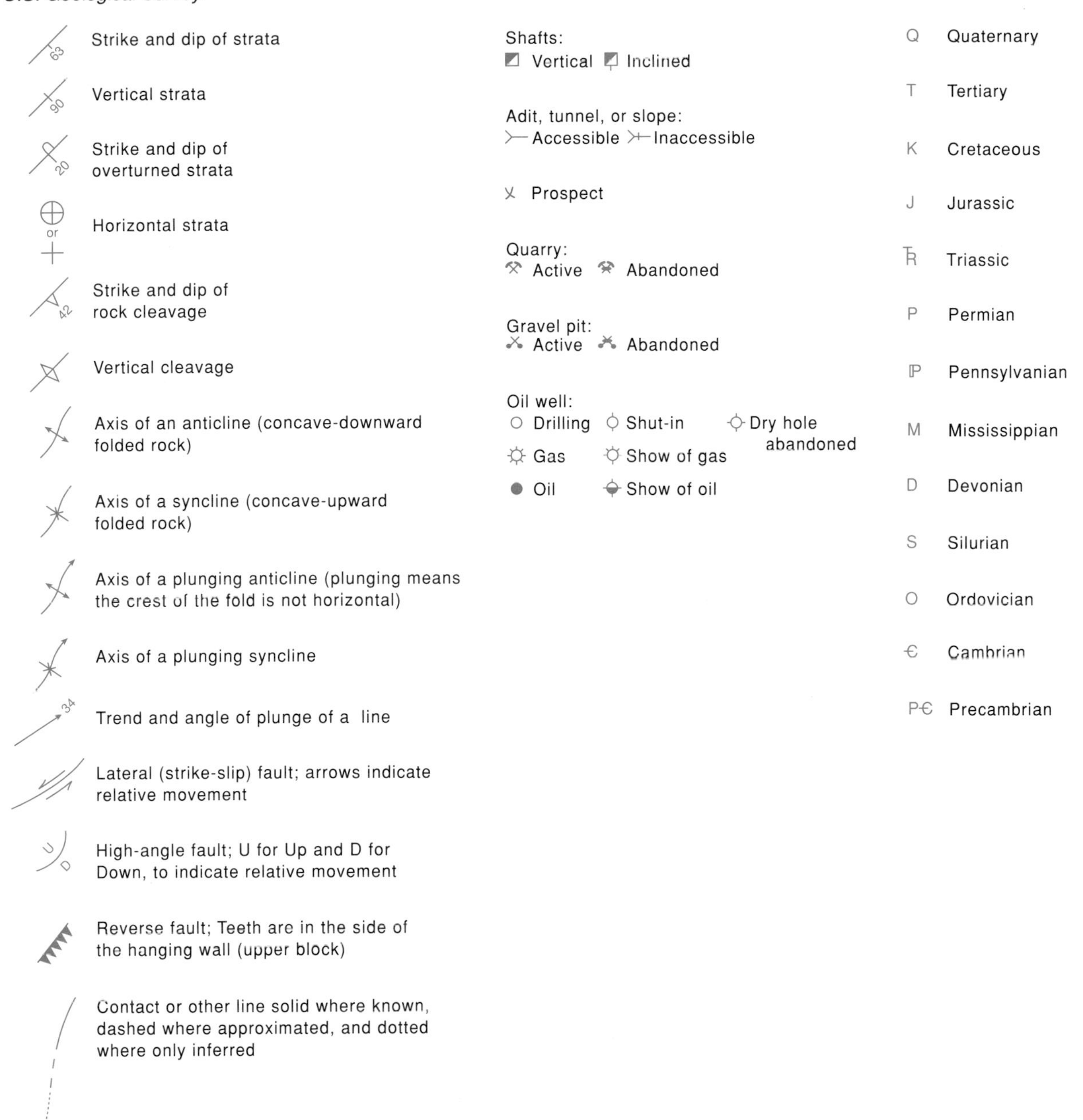


become cavernous underground, posing a danger to buildings constructed directly above? These are some of the reasons why environmental geologists must be able to visualize the materials buried beneath the ground surface (figure 6.4). The result of this visualization is a *geologic cross-section.*

Some of the data used in constructing a cross-section, such as strike and dip, come from observations made at ground level. Others are obtained from wells drilled to find water, petroleum, and natural gas; from mines excavated for mineral resources; or from tunnels excavated to enlarge transportation networks. Wells drilled for the petroleum industry are useful because they bring rock chips to the surface, and because they are "logged" in some form. Wireline logs are zigzag lines on a strip of paper that indicate how subsurface rocks respond to electrical or nuclear impulses. The character of the response allows geologists to interpret lithology, water content, and other useful properties of the rock. Another technique for analyzing subsurface rocks is to use explosives to send energy impulses into the ground. Contacts between rock layers reflect some of the energy upward, permitting determination of the type of rock and its dip. The dip of the layer at the surface is not necessarily the same as its dip below the surface.

Constructing a geologic cross-section is very much like constructing a topographic profile:

1. Orient the cross-section line approximately normal to the major geologic trends in the area, that is, normal to the strike of beds, folds, and faults.
2. Construct a topographic profile along the cross-section line.
3. Use tick marks to transfer the geologic contacts along the line to the topographic profile, as was done for elevations in the topographic profile.
4. Where dips are known, extend formational contacts and fault planes down into the profile, using a protractor to measure angles.
5. Connect at depth all outcrops of the same sedimentary rock unit, using a reasonable extension of the surface dips. Most units maintain a fairly constant thickness. If the cross-section includes a fault, be aware that rock units will not match up exactly across the fault. If one side of the fault has moved up or down, the formation contacts will be displaced across the fault surface.
6. Label the cross-section, using the appropriate letter symbols from the map. Also place arrows alongside any faults to show relative displacement.

## *Principles of Geologic Map Interpretation*

After a geologic cross-section is constructed, it must be interpreted to answer questions such as the relative ages of the rock layers and faults or the ages of unconformities. The geologic principles on which these interpretations are made are the following:

1. *Original horizontality.* Layers of sedimentary rock were originally deposited as horizontal layers of sediment. If these rocks are now dipping, they must have been tilted after deposition.

**Figure 6.3**

Strike and dip of a limestone layer resting on granite. Strike is east-west and dip is to the south at about 30°.

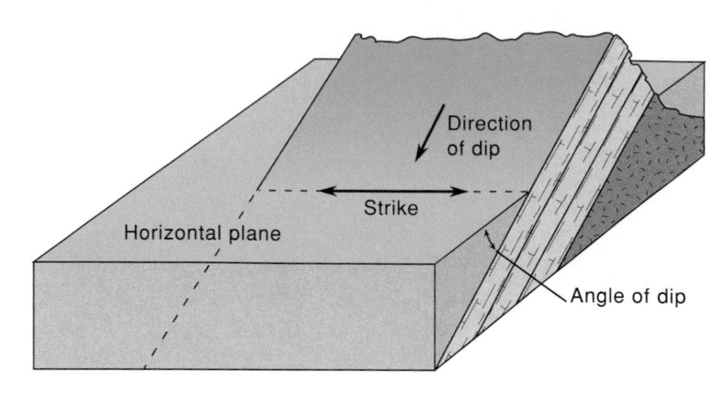


2. *Superposition.* In an undisturbed sequence of sedimentary rocks, the oldest rock is at the bottom. Each succeeding layer is younger than the layer below it.
3. *Cross-cutting relationships.* A feature that cuts across a layer of sedimentary rock must be younger than the layer being cut. Thus, a fault is younger than the beds it displaces, and an igneous intrusion is younger than the rocks it intrudes. An unconformity must be younger than the rocks it cuts. A layer must be present before it can be cut, intruded, or eroded.
4. *Inclusions.* An inclusion is a fragment of a pre-existing body of rock, present within a younger rock. For example, a magmatic intrusion might contain a piece of sedimentary rock broken off during the ascent of the magma. The intrusion must be younger than the piece of sedimentary rock. Commonly the sedimentary rock layer directly above an unconformity surface contains fragments from rock layers below the unconformity. An unconformity is an ancient erosional surface—an ancient ground surface—and rock fragments were lying on the surface when the sedimentary layer above the unconformity was deposited. Inclusions, although often visible in outcrops, are not normally indicated on geologic maps.

## Figure 6.4

Patterns shown by dipping beds exposed in a stream valley, showing the "rule of *V*s."

If beds are vertical, contacts will be straight, no matter what slopes they cross.

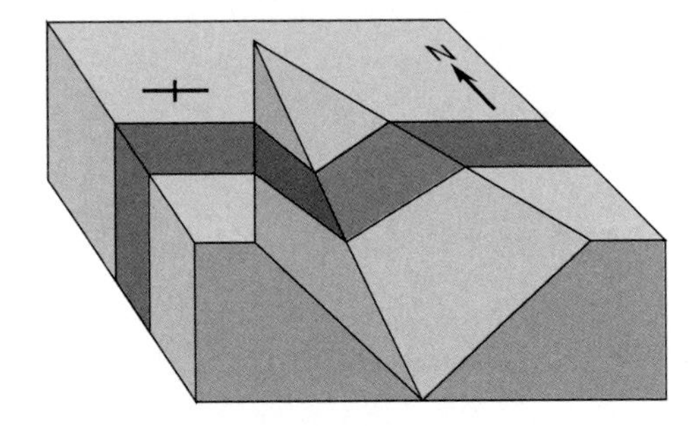


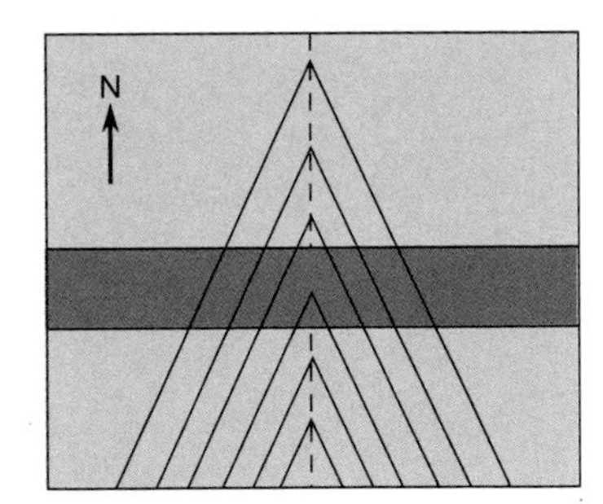


If beds are horizontal, contacts will make a V pointing upstream, parallel to the contours.

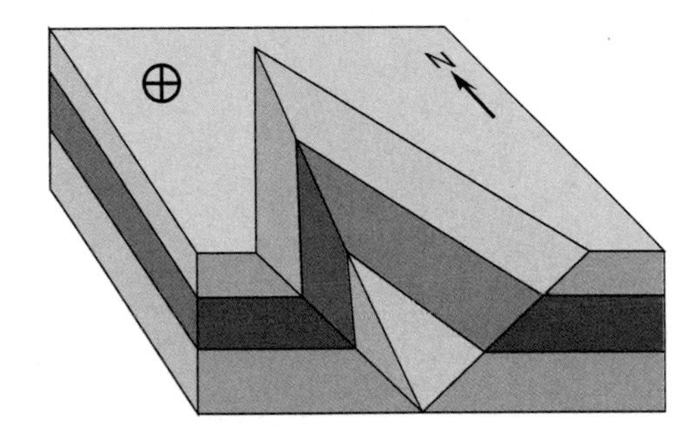


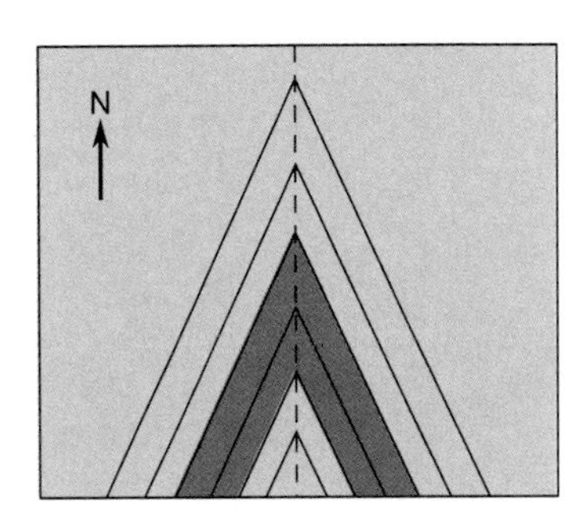


If beds dip upstream, contacts will make a V pointing upstream but crossing the contours.

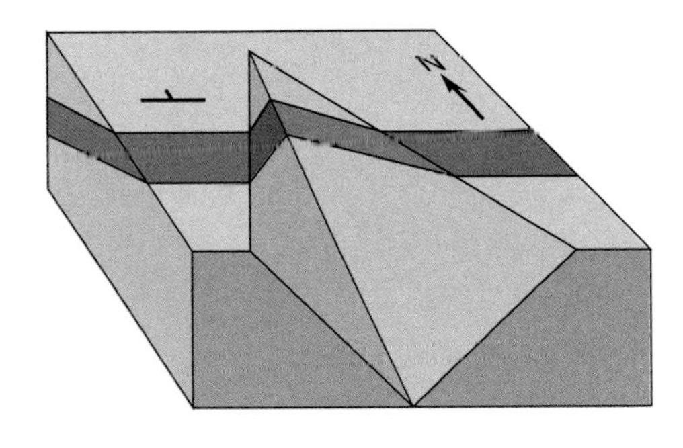


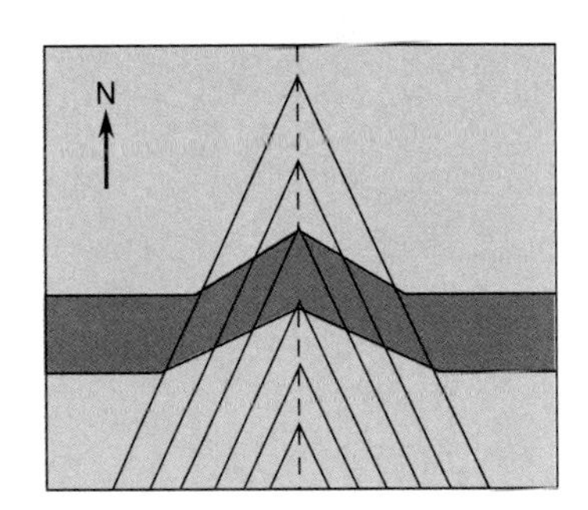


If beds dip downstream steeper than channel, contacts will make a V pointing downstream.

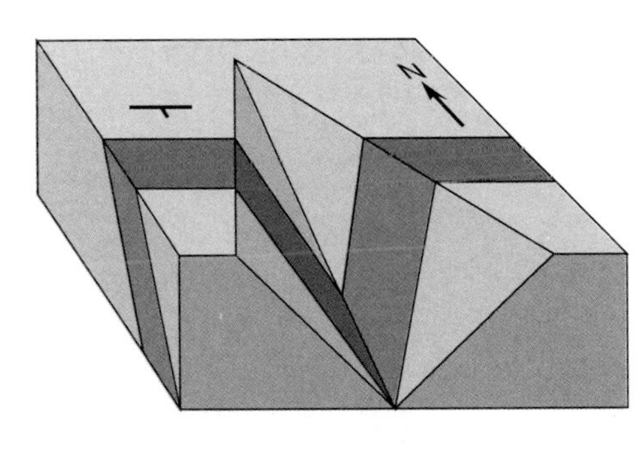


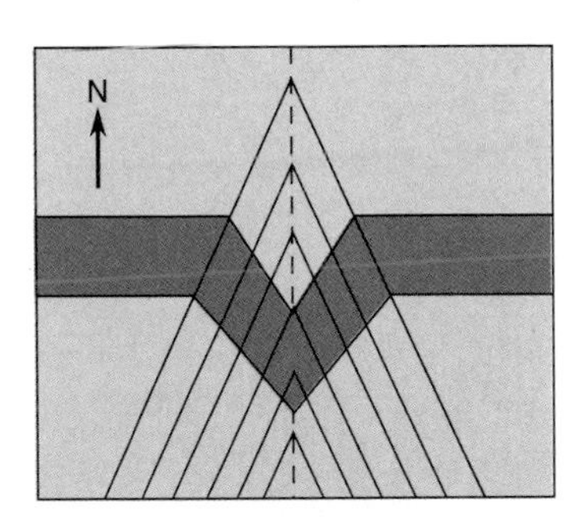


If beds dip downstream shallower than channel, contacts will make an acute V pointing upstream.

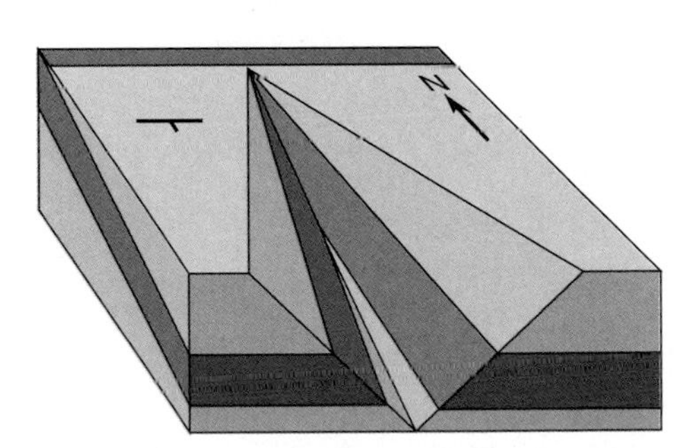


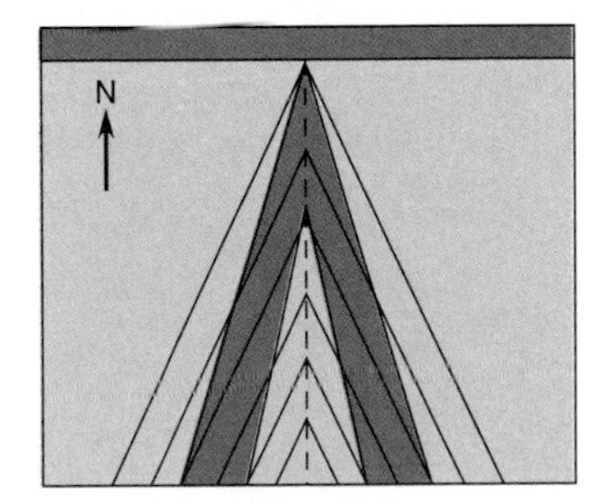

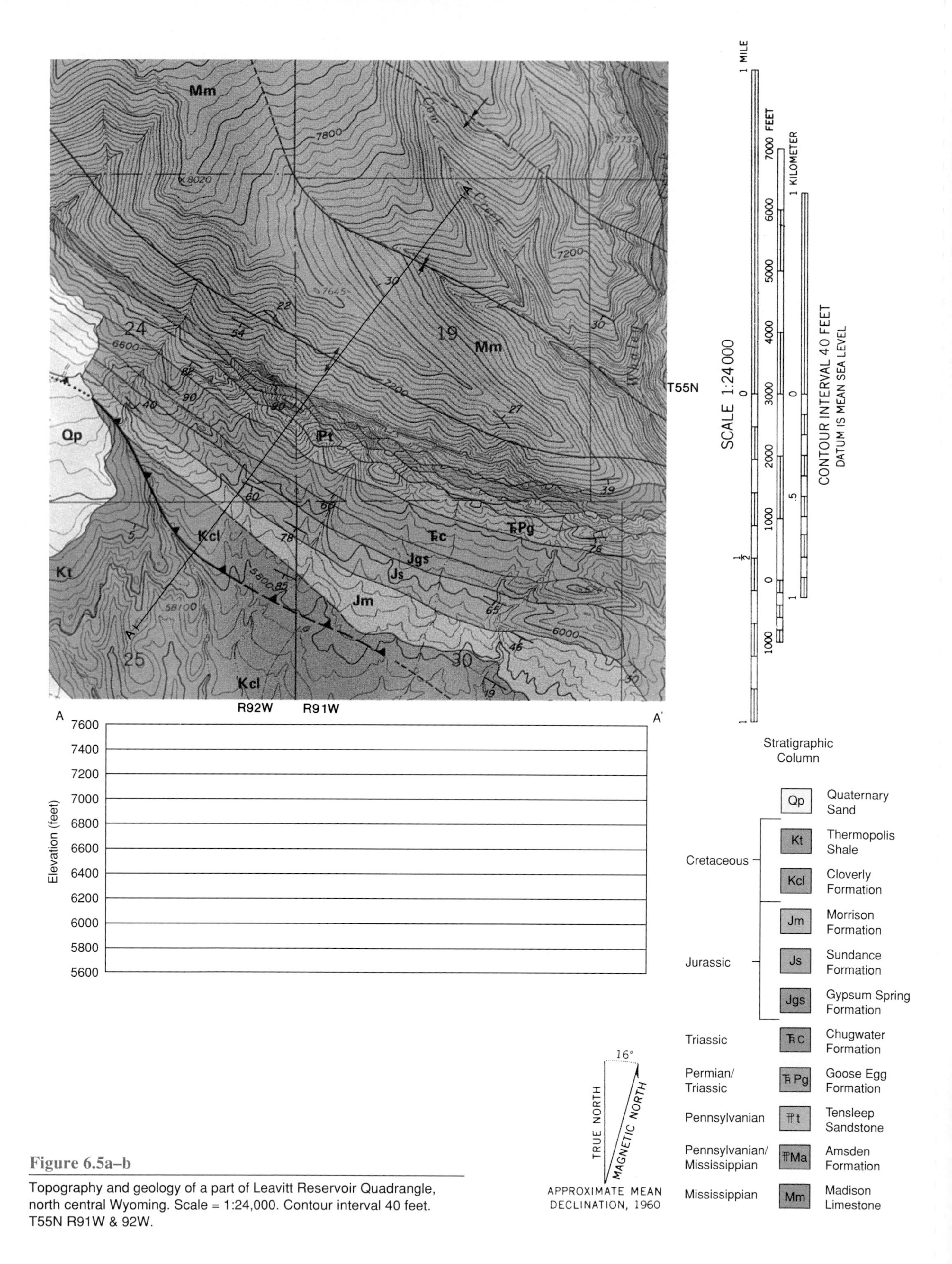


**Figure 6.5a–b**

Topography and geology of a part of Leavitt Reservoir Quadrangle, north central Wyoming. Scale = 1:24,000. Contour interval 40 feet. T55N R91W & 92W.

## Problems

1. Construct a topographic and geologic cross-section along line A-A′ on the Leavitt Reservoir Quadrangle (figure 6.5).
2. The map symbols indicate the presence of a syncline with its axis trending NW–SE through the northern half of section 19. Would you know about the existence of this structure if it were not indicated on the map? Explain.
3. Does your answer apply also to the anticlinal axis in the southern half of the section? Would you be able to draw in the approximate position of the axis if the line were not present on the map?
4. Could you determine the direction of dip of the beds in the SE 1/4 of section 24 if no strike and dip symbols were present? If so, how?
5. Do you think an unconformity can develop at the bottom of the sea? Explain.
6. In studying an unconformity, how might you try to determine how long it took the surface to develop, that is, the amount of time the surface represents?
7. Locate an unconformity surface on the map.

## Further Reading/References


Barnes, J., 1991, *Basic Geological Mapping,* 2nd ed. New York, John Wiley & Sons, 118 pp.

Bolton, T., 1989, *Geological Maps.* New York, Cambridge University Press, 144 pp.

Boulter, C. A., 1989, *Four Dimensional Analysis of Geological Maps.* New York, John Wiley & Sons, 296 pp.

Butler, B. C. M., and Bell, J. D., 1988, *Interpretation of Geological Maps.* New York, John Wiley & Sons, 236 pp.

Maltman, A., 1990, *Geological Maps: An Introduction.* New York, Van Nostrand Reinhold, 184 pp.

McCall, J., and Marker, B. (eds.), 1989, *Earth Science Mapping for Planning, Development and Conservation.* Boston, Graham & Trotman, 268 pp.

exercise

# SEVEN

## Seismic Risk and Earthquakes

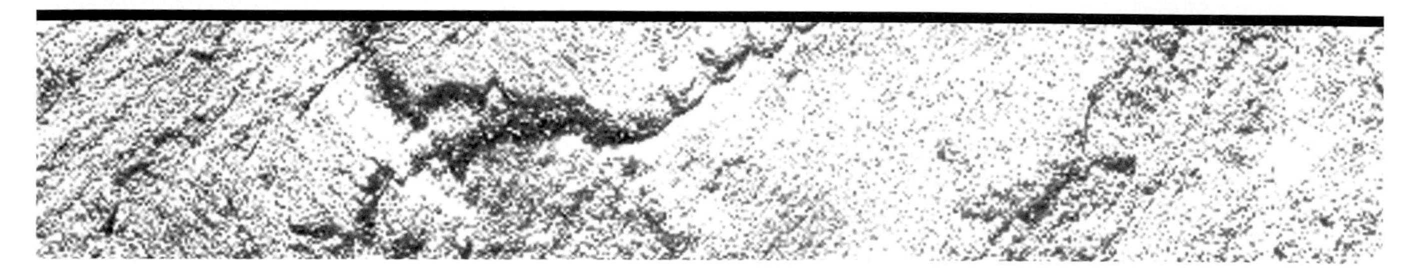

For most of us, the word "earthquake" triggers visions of violently trembling ground, landslides, collapsing buildings, fires, and great loss of life. These images come from the popular press and commercial films, which give the impression that earthquakes are infrequent, but catastrophic when they do occur. In actuality, about 350,000 earthquakes occur each year, and fewer than 1,000 (<0.3%) cause noticeable damage or loss of life. Furthermore, 72% of the damaging quakes are concentrated in a narrow band around the periphery of the Pacific Ocean. Most of the rest occur in the mountain belt that extends from the Alps in southern France eastward through the Middle East to the Himalayan range in northern India and southern China. In the United States the area of most concern is the California coast, which has a high population density, two of the nation's largest cities (Los Angeles and San Francisco), and considerable topographic relief on unstable slopes. These three factors are a prescription for disaster in an earthquake-prone region.

Most earthquakes result from shock waves created by the sudden release of slowly accumulated stress in rigid bedrock. This release of stress causes blocks of rock to move along fractures called faults. The rock then becomes offset on either side of the fault. The sliding motion relieves the stress for a while, but as stress continues to accumulate, another sudden movement parallel to the fault surface is inevitable. This situation exists all along the San Andreas fault (actually a fault zone with many large and small subparallel faults), which runs just east of Los Angeles and slightly west of San Francisco. The slice of the Earth's crust west of the San Andreas fault has been moving northward for about 40–50 million years at an average rate of roughly 1 cm/yr. The human problem is that this movement does not occur by continuous sliding at that imperceptible rate; instead, strain energy is stored continuously, and then relieved sporadically—and unpredictably—by rapid movements of several centimeters or more.

### *Magnitude and Intensity*

The *magnitude* of an earthquake refers to the amount of energy released, based on a scale devised by Charles F. Richter. This energy release is measured by the amount of ground displacement or shaking the earthquake produces. The Richter scale is logarithmic, which means that an earthquake of magnitude 5 causes ten times as much ground movement as one of magnitude 4—and a hundred times as much movement as one of magnitude 3. At increased magnitudes the amount of energy released rises even faster, by a factor of more than 30 for each unit of magnitude (table 7.1A). The largest recorded earthquakes had magnitudes of about 8.9.

An alternate way to describe the size of an earthquake is by its intensity, a measure of the effect of the quake on people and structures. Intensity varies considerably because of factors such as local geologic conditions, quality of construction, and distance from the epicenter of the earthquake. The epicenter is the place on the ground surface directly above the spot where the rock ruptures. The magnitude of any particular earthquake is a constant; its intensity is a variable. Many intensity scales are in use, but in the United States the most widely used is the Modified Mercalli Scale (table 7.1B).

The amount of property damage from an earthquake of a particular magnitude at a given distance from the epicenter depends mostly on the character of the underlying sediment or bedrock, as was clearly illustrated in the famous

TABLE 7.1a

Frequency of Earthquakes of Various Magnitudes on the Richter Scale and the Amount of Energy Released.

| Description | Magnitude | Number per year | Approximate energy released (ergs) |
|---|---|---|---|
| Great earthquake | over 8 | 1 to 2 | over $5.8 \times 10^{23}$ |
| Major earthquake | 7–7.9 | 18 | $2–42 \times 10^{22}$ |
| Destructive earthquake | 6–6.9 | 120 | $8–150 \times 10^{20}$ |
| Damaging earthquake | 5–5.9 | 800 | $3–55 \times 10^{19}$ |
| Minor earthquake | 4–4.9 | 6,200 | $1–20 \times 10^{18}$ |
| Smallest usually felt | 3–3.9 | 49,000 | $4–72 \times 10^{16}$ |
| Detected but not felt | 2–2.9 | 300,000 | $1–26 \times 10^{15}$ |

*Source: Data from B. Gutenberg in* Earth, *2d ed by Frank Press and Ray Siever, 1978, W. H. Freeman and Company.*

TABLE 7.1b

Modified Mercalli Intensity Scale (Abridged).

| Intensity | Description |
|---|---|
| I | Not felt. |
| II | Felt by persons at rest on upper floors. |
| III | Felt indoors—hanging objects swing. Vibration like passing of light trucks. |
| IV | Vibration like passing of heavy trucks. Standing automobiles rock. Windows, dishes, and doors rattle; wooden walls or frames may creak. |
| V | Felt outdoors. Sleepers wakened. Liquids disturbed, some spilled; small objects may be moved or upset; doors swing; shutters and pictures move. |
| VI | Felt by all; many frightened. People walk unsteadily; windows and dishes broken; objects knocked off shelves, pictures off walls. Furniture moved or overturned; weak plaster cracked. Small bells ring. Trees and bushes shaken. |
| VII | Difficult to stand. Furniture broken. Damage to weak materials, such as adobe; some cracking of ordinary masonry. Fall of plaster, loose bricks, and tile. Waves on ponds; water muddy; small slides along sand or gravel banks. Large bells ring. |
| VIII | Steering of automobiles affected. Damage to and partial collapse of ordinary masonry. Fall of chimneys, towers. Frame houses moved on foundations if not bolted down. Changes in flow of springs and wells. |
| IX | General panic. Frame structures shifted off foundations if not bolted down; frames cracked. Serious damage even to partially reinforced masonry. Underground pipes broken; reservoirs damaged. Conspicuous cracks in ground. |
| X | Most masonry and frame structures destroyed with their foundations. Serious damage to dams and dikes; large landslides. Rails bent slightly. |
| XI | Rails bent greatly. Underground pipelines out of service. |
| XII | Damage nearly total. Large rock masses shifted; objects thrown into the air. |

**Figure 7.1a**

Generalized geologic map of the upper part of the San Francisco Peninsula, California.
*Source: Borcherdt, 1975, p. 4, U.S. Geological Survey.*

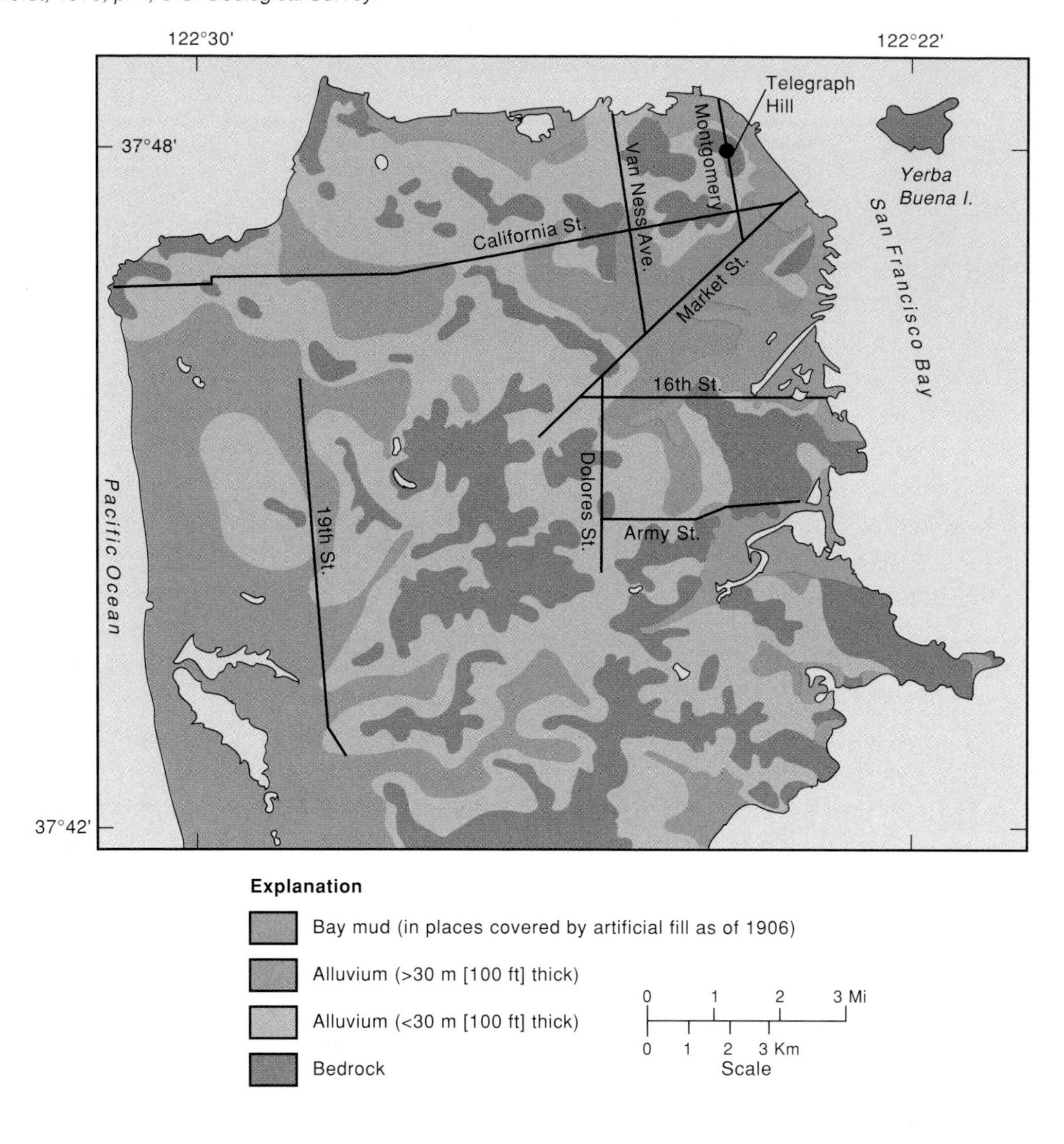


San Francisco earthquake of April 18, 1906 (figure 7.1). In the Telegraph Hill area, where bedrock is exposed at the surface, the effects of the earthquake were "weak" (figure 7.2), with occasional falling chimneys and damage to plaster, partitions, and plumbing. But about 1,000 feet from Telegraph Hill, in an area underlain by artificial fill and water-saturated mud, the effects of the quake were "violent," with widespread collapsing of brick and frame structures. These differential effects demonstrate the need for weighing geologic conditions when zoning earthquake-prone areas. Zoning should also consider whether the terrain is steep or flat and evaluate the structural characteristics of man-made features such as buildings, dams, and bridges.

## *Man-Induced Earthquakes*

Earthquakes can be caused by human activities as well as by natural causes, as exemplified by an earthquake swarm that occurred near Denver, Colorado, between 1962 and 1965 (Evans, 1966). Since 1942, chemical-warfare products had been manufactured on a large scale at the Rocky Mountain Arsenal, about 10 miles northeast of Denver. One by-product of this operation was contaminated wastewater that, until 1961, was disposed of by evaporating it from dirt reservoirs. After the wastewater was found to be contaminating the local groundwater supply and endangering crops, an injection well was drilled so the water could be disposed of below the reach of surface processes. The well was drilled through 11,895 feet of sedimentary rocks into the gneissic basement, to a total depth of 12,045 feet, and fluid was pumped down it at rates as high as 9 million gallons

**Figure 7.1b**

Distribution of apparent intensities of the 1906 earthquake in San Francisco, California. Note the strong correlation between underlying type of sediment and rock and the damage sustained (intensities).

*Source: Borcherdt, 1975, p. 3, U.S. Geological Survey.*

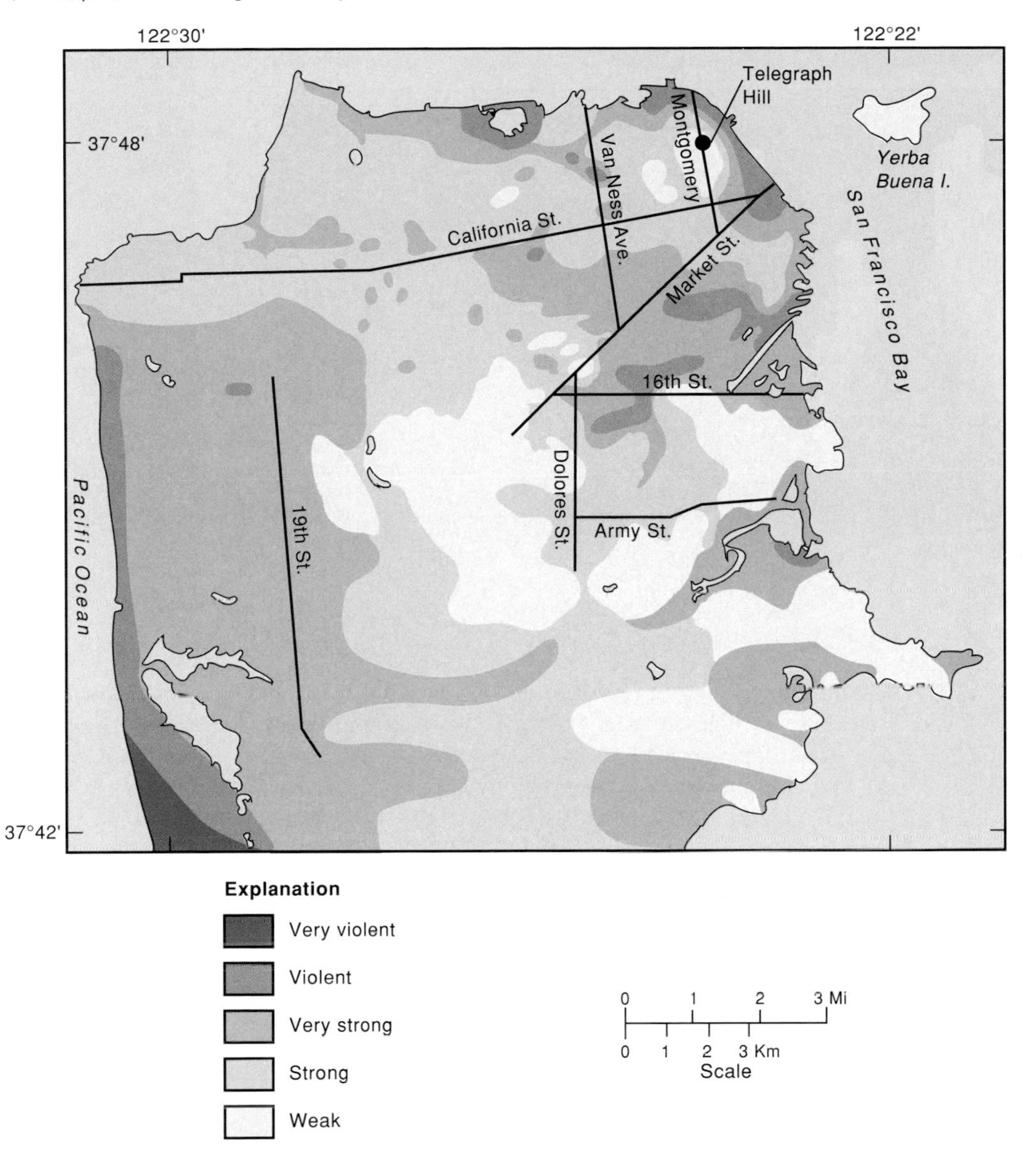


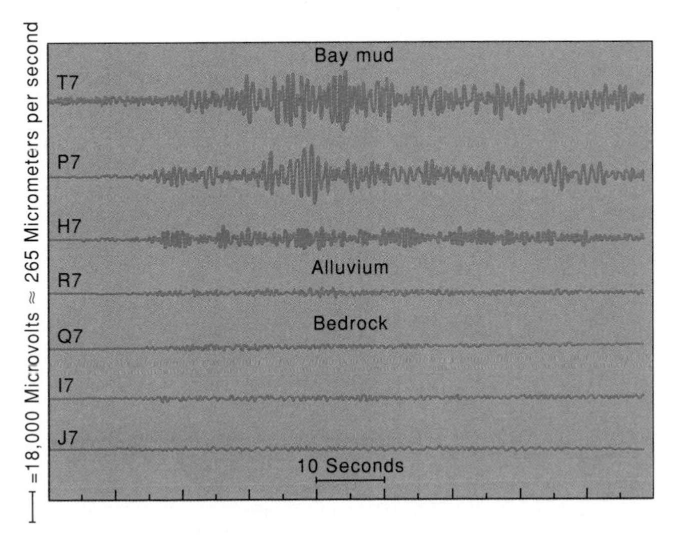


**Figure 7.2**

Recording of horizontal ground motion generated by an underground nuclear explosion in Nevada on sediment and bedrock in San Francisco. The effects are the same as those of an earthquake. The vertical scale shows the amount of ground motion; the horizontal scale shows elapsed time. Results are shown for three sites on recently deposited plastic mud, tens of meters thick, that contains more than 50% water; for one site underlain by clay, sand, and gravel hundreds of meters thick that contains less than 40% water; and for three sites underlain by semilithified and lithified sedimentary rocks of varied thicknesses.

*Source: Borcherdt, 1975, p. 54, U.S. Geological Survey.*

**Figure 7.3**

(a) Number of earthquakes per month recorded in the Denver area. (b) Monthly volume of contaminated wastewater injected into the Arsenal well.

*Source: Data from Evans,* Mountain Geologist, *Volume 27, page 27, 1966, Rocky Mountain Association of Geologists, Denver, CO.*

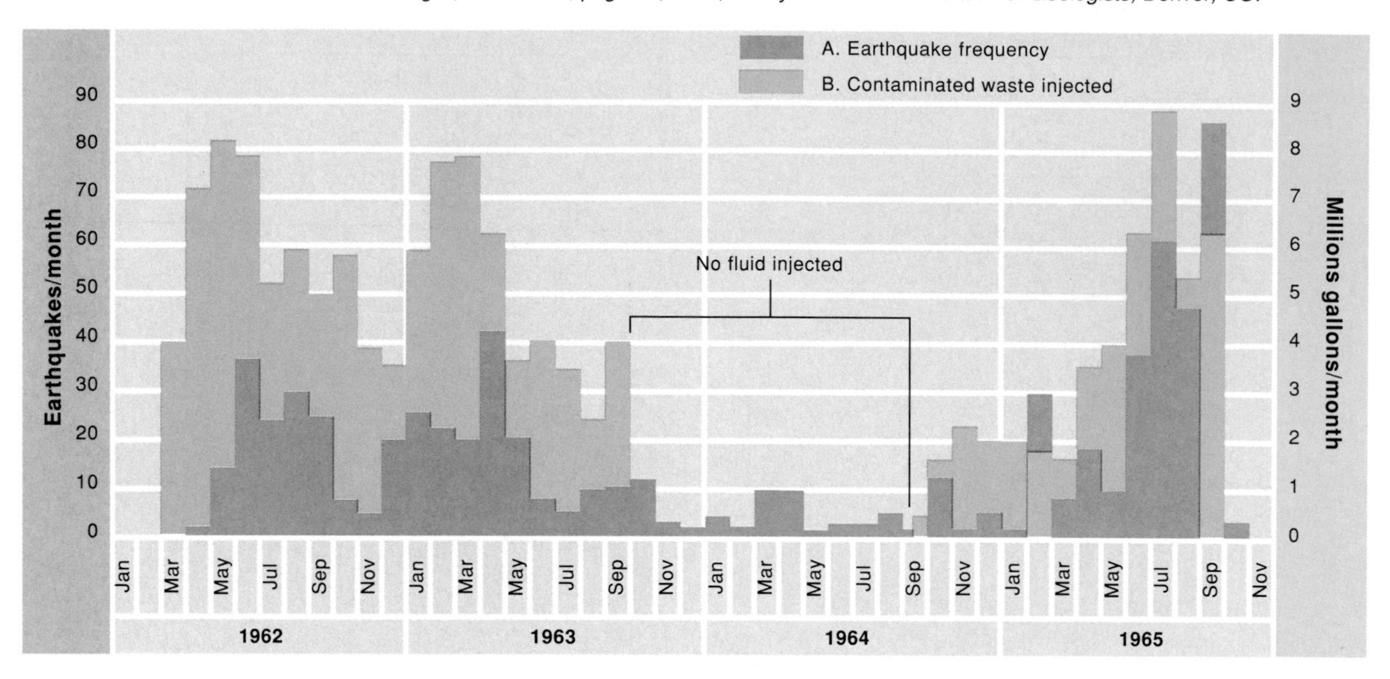


per month, to a total of 150 million gallons (figure 7.3). The result of this disposal method was that between April 1962 and September 1965, 710 earthquakes with magnitudes of 0.7 to 4.3 occurred with epicenters in the Denver area—although only one earthquake had been recorded there before, in 1892. Clearly, a strong correlation existed between the fluid injection and earthquake frequency; the injected fluid decreased the friction between fracture surfaces, triggering fault movement. Despite the federal government's repeated denials of culpability, the disposal well was eventually shut down. The earthquakes attributable to this disposal well are simply another example of a harmful interaction between humans and their natural environment.

## *Earthquake Prediction*

The problem of earthquake prediction is currently receiving considerable attention in Japan, the former Soviet Union, China, and the United States, all of which have suffered significant property damage and loss of life from large earthquakes in recent years. Although a few successful predictions have been made, no reliable method of short-range prediction yet exists. The development of such a method seems many decades in the future and, even then, it is questionable whether any method will be accurate enough in view of economic incentives and human behavior. For example, suppose seismologists (scientists who study earthquakes) announce that a major earthquake is "likely" to occur in May or June "in the vicinity of" Oakland, about 10 miles east of downtown San Francisco. How useful is this information? Is everyone within a 10- to 20-mile radius of Oakland likely to leave their houses for two months or more? Will offices and industries shut their doors during this period? If the "likely" quake fails to occur, will lawsuits be filed by people who have suffered financially because of the evacuation? Conversely, can seismologists be held responsible for failing to predict an earthquake that *does* occur? Should state legislatures pass a Good Samaritan law for seismologists, such as some states already have adopted to protect people who assist accident victims? Such questions will become more significant in the future, as predictions of the locations and timing of earthquakes become increasingly more accurate.

## Problems

1. The Richter scale of magnitude is open-ended, with no upper limit. However, no earthquake has been recorded with a magnitude greater than 8.9 and, as is evident in table 7.1, only a trivial number of earthquakes have magnitudes of 8 or more. Explain these observations.
2. Figure 7.3 shows a one-year period during which no fluid was injected into the disposal well, but earthquakes occurred without interruption. How might you explain this? What might explain the less-than-perfect correlation between the amount of water injected each month and the number of earthquakes?
3. Figure 7.4 shows the numerical data for earthquake intensities during the San Fernando, California, shock (magnitude 6.4) of February 9, 1971.
    a. Construct an isointensity contour map at a unit interval for the data and give plausible reasons for the regularities and irregularities in the shapes of the contours.

Figure 7.4

Intensity distribution map of the San Fernando, California, earthquake of February 9, 1971.
*Source: Blair et al, 1971, p. 17, U.S. Geological Survey.*

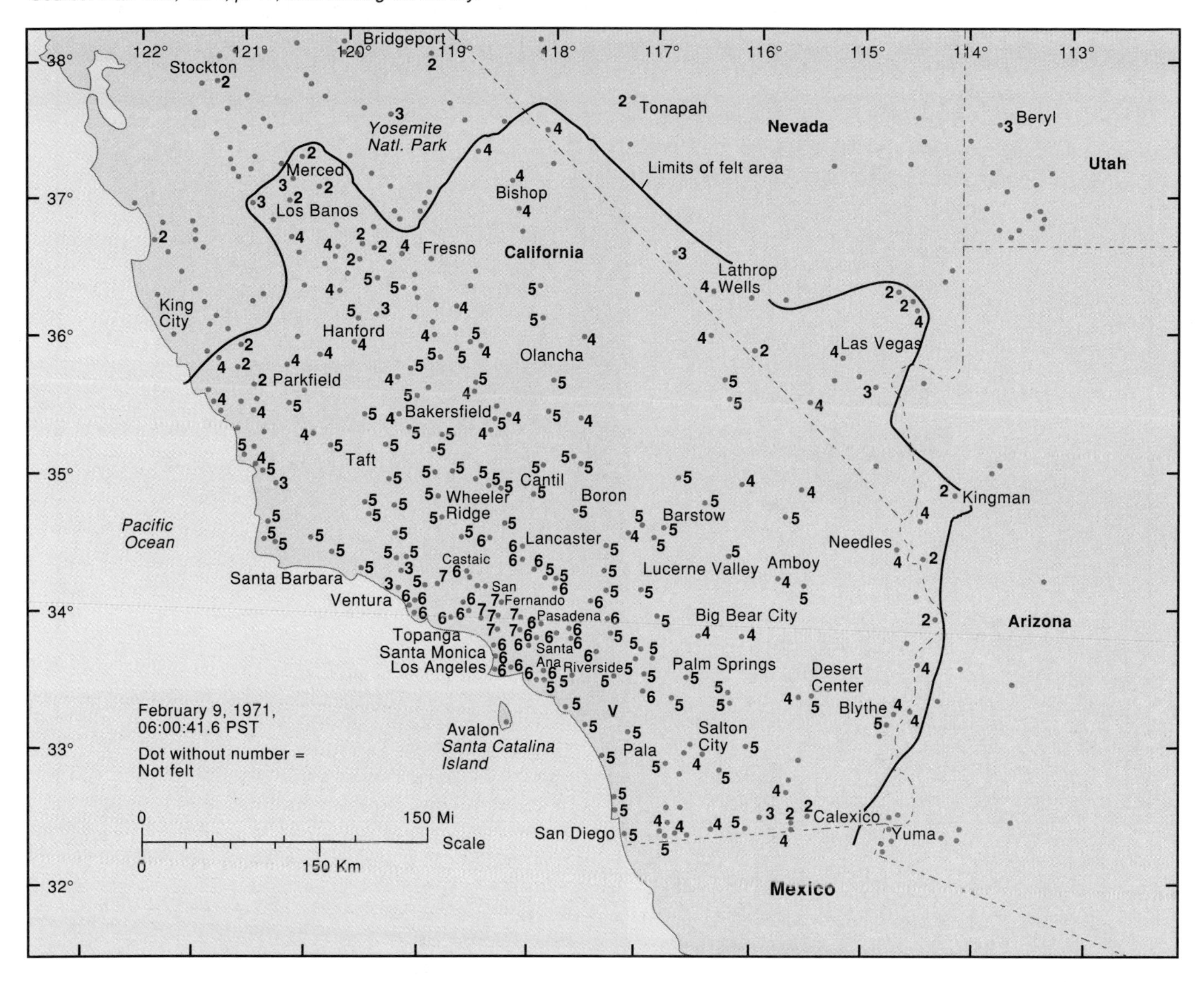


b. Construct an X–Y graph of earthquake intensity versus distance between San Fernando, where the maximum intensity recorded was 11, to Las Vegas. Interpret the shape of the curve.

4. What intensity would you assign to the damage described in the text for Telegraph Hill in the 1906 San Francisco earthquake? What intensity in the area 1,000 ft from Telegraph Hill?

5. The description of damage from the 1906 San Francisco earthquake summarized in the text cites only damage to buildings caused by shaking-induced structural failures. What other types of effects might have occurred in the affected area?

6. The San Andreas fault trends N–S up the San Francisco peninsula (perhaps along 19th Street on figure 7.1), and the Hayward Fault runs parallel to it and 20 miles to the east, near Oakland. Seismologists have projected a 67% probability that an earthquake of magnitude 7 or more will occur along one of these faults within the next 30 years. What, if anything, should the city councils of potentially affected areas do with this information? What can insurance companies do with it? What should residents of these areas do?

7. Figure 7.5 is a generalized lithologic, isopach, and topographic map of southern Sonoma County, California, just north of San Francisco. Movement along the San Andreas fault in 1906 created offsets of as much as 12 feet in Sonoma County. Seven other potentially active faults have been identified in the area, all trending NW-SE, parallel to the bedrock outcrop on figure 7.5. Describe how you would approach the problem of creating a seismic-risk analysis report for the residents of the area. What types of information would you like to have that are not given on the map?

**Figure 7.5**

Rock and sediment distributions and thicknesses in southern Sonoma County, California.

*From R. W. Greensfelder, 1980, California Division of Mines and Geology* Special Report 120, *plate 1B. Used with permission.*

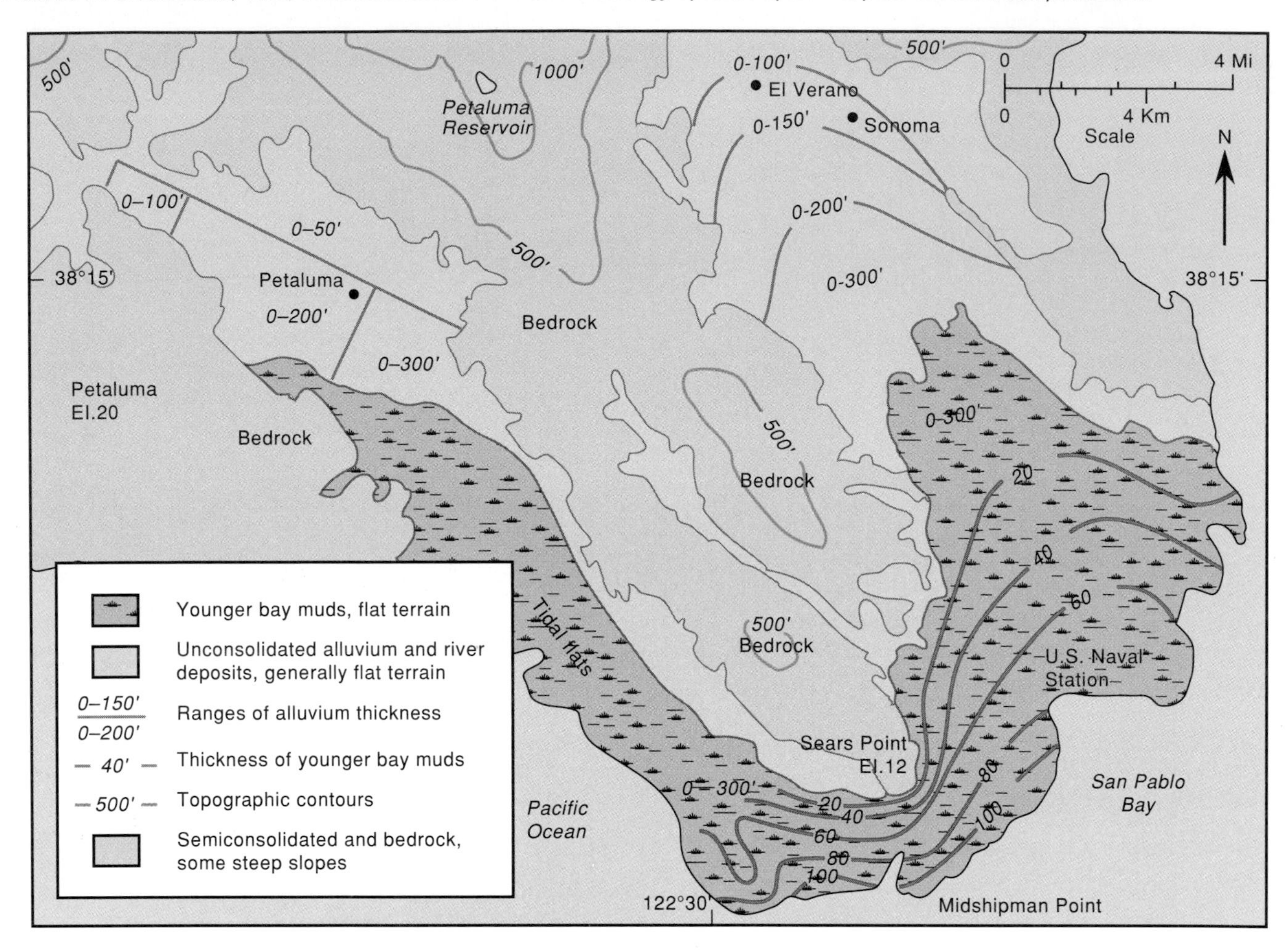


## Further Reading/References


Blair, M. L., and W. E. Spangle, 1979, *Seismic Safety and Land-use Planning—Selected Examples from California.* U.S. Geological Survey Professional Paper 941-B, 82 pp.

Borcherdt, R. D. (ed.), 1975, *Studies for Seismic Zonation of the San Francisco Bay Region.* U.S. Geological Survey Professional Paper 941-A, 102 pp.

California Seismic Safety Commission, 1992, *The Homeowner's Guide to Earthquake Safety.* Seismic Safety Commission # 92–02, 28 pp.

Dieterich, J. H., et al., 1990, *Probabilities of Large Earthquakes in the San Francisco Bay Region, California.* U.S. Geological Survey Circular 1053, 51 pp.

Evans, D. M., 1966, The Denver area earthquakes and the Rocky Mountain Arsenal disposal well: *The Mountain Geologist,* v. 3, no. 1, pp. 23–36.

Moran, D. E., Slosson, J. E., Stone, R. O., and Yelverton, C. A. (eds.), 1973, *Geology, Seismicity, and Environmental Impact.* Los Angeles, University Publishers, 444 pp.

Nicholson, C., and Wesson, R. L., 1990, *Earthquake Hazard Associated with Deep Well Injection—a Report to the U.S. Environmental Protection Agency.* U.S. Geological Survey Bulletin 1951, 74 pp.

Plafker, G., and Galloway, J. P. (eds.), 1989, *Lessons Learned from the Loma Prieta, California, Earthquake of October 17, 1989.* U.S. Geological Survey Circular 1045, 48 pp.

Reiter, L., 1991, *Earthquake Hazard Analysis.* New York, Columbia University Press, 254 pp.

Schmidt, R. G., 1986, *Geology, Earthquake Hazards, and Land Use in the Helena Area, Montana—a Review.* U.S. Geological Survey Professional Paper 1316, 64 pp.

Steinbrugge, K. V., and Algermissen, S. T., 1990, *Earthquake Losses to Single-family Dwellings: California Experience.* U.S. Geological Survey Bulletin 1939-A, 65 pp.

Yanev, P. I., 1991, *Peace of Mind in Earthquake Country.* San Francisco, Chronical Books, 224 pp.

Ziony, J. I. (ed.), 1985, *Evaluating Earthquake Hazards in the Los Angeles Region—an Earth-science Perspective.* U.S. Geological Survey Professional Paper 1360, 505 pp.

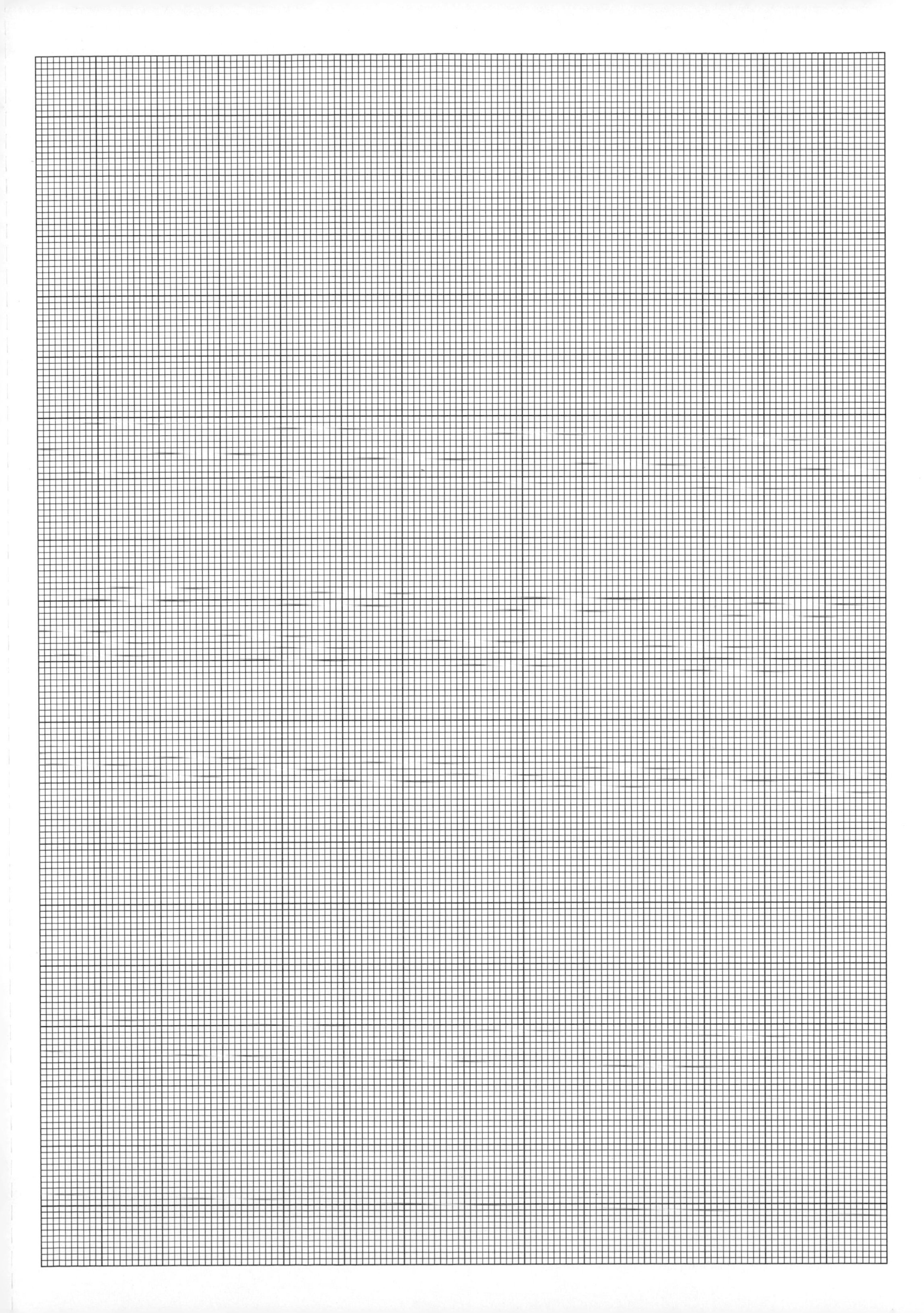

exercise

# EIGHT

# Volcanoes and Eruptions

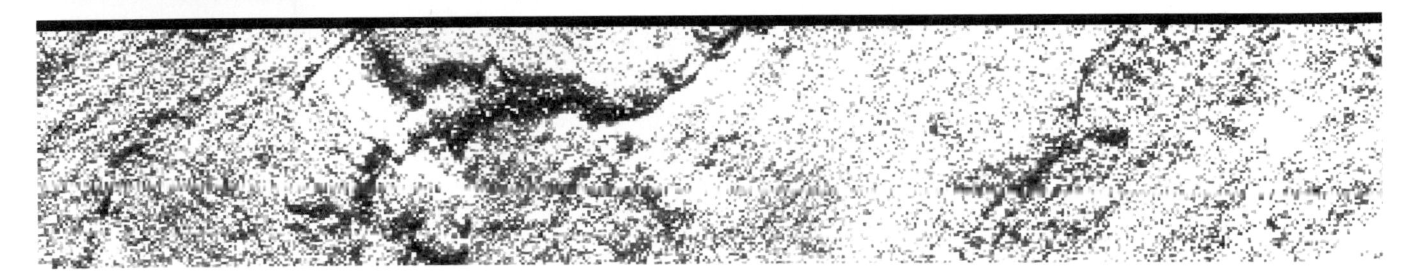

Volcanic eruptions occur infrequently and cause low annual damage compared to other hazards, such as earthquakes, floods, and ground failures (subsidence, sinkholes, landslides). In the United States, for example, the annual economic loss from volcanic eruptions is at least an order of magnitude less than the loss ($0.6 \times 10^9$ 1980 dollars) caused by earthquakes, which in turn is nearly an order of magnitude less than the loss associated with floods and ground failures. Nevertheless, for those who live in volcanic regions, the severity of the hazards caused by potential eruptions makes them an important consideration in land-use decisions.

Because of this fact, scientists have attempted to compile lists of high-risk volcanoes. The rating criteria for such lists include some or all of the following factors: (1) frequencies, sites, and nature of recorded historical eruptions (figure 8.1); (2) information on recent prehistoric eruptions as inferred from mapping and dating studies (figure 8.1); (3) known ground deformation and/or seismic events ("earthquake swarms"); (4) the nature of eruptive products, possible indicators of explosive potential; and (5) various demographic determinants, such as population density, property at risk, and fatalities and/or evacuations resulting from recorded historical volcanic disasters or crises. Unfortunately, all such lists are incomplete because geological and geophysical data for many volcanoes are inadequate. Many eruptions have occurred from volcanoes not previously considered high-risk.

Despite these uncertainties, volcanoes are commonly classified as *active, dormant,* or *extinct.* An active volcano is one that has erupted within recent history. If the volcano has not erupted within historic times—perhaps 5,000 years—but is fresh-looking and not significantly eroded, it is considered dormant, with the potential to become active again. A volcano is considered extinct if it shows significant erosion of its crest and flanks and has no recent eruptive history.

## *Types of Eruptions*

Eruptions can be either explosive or nonexplosive. Explosive eruptions occur if the magma moving up the neck of the volcano is very viscous, so that most of the gases present in the magma remain dissolved. As the magma nears the surface the pressure decreases; increasing amounts of gas come out of solution, and the gas bubbles increase in size. Eventually the pressure becomes great enough to make the gases burst out of the viscous liquid, creating the explosion that hurls great masses of magma and semiconsolidated volcanic debris from the throat of the volcano. Groundwater heated by the rising magma is also ejected in the explosion; in fact, the bulk of the fluids emitted from volcanoes is $H_2O$ liquid and vapor.

The solid material ejected from a volcano is called *tephra,* and the layered sediment that accumulates around the eruption site is termed *pyroclastic sediment,* a deposit on the borderline between igneous and sedimentary. The height and lateral distance that fragments attain depend on ejection force, the size of the fragments, and wind velocity. Fragments larger than 60 mm in diameter are called bombs (round because they solidify during flight) or blocks (angular fragments); those between 60 mm and 2 mm are lapilli; those smaller than 2 mm are ash or dust. Tephra can endanger life and property at considerable distances from the volcano by forming a blanket over the ground surface and contaminating the air with abrasive particles and corrosive acids. Close to a volcano, people can be injured or killed by

**Figure 8.1**

Map of Mauna Loa showing the surface distribution of lava flows in five different age categories. The notation "ka" stands for thousands of years before the year 1950. Thus, 0.75 ka = 750 years; 1.5 ka = 1,500 years; and 4.0 ka = 4,000 years.

*Source: J. P. Lockwood and P. W. Lipman, 1987, Holocene Eruption History of Mauna Loa Volcano, Chapter 18 in R. W. Decker et al., editors, Volcanism in Hawaii, U.S. Geological Survey* Prof. Paper 1350.

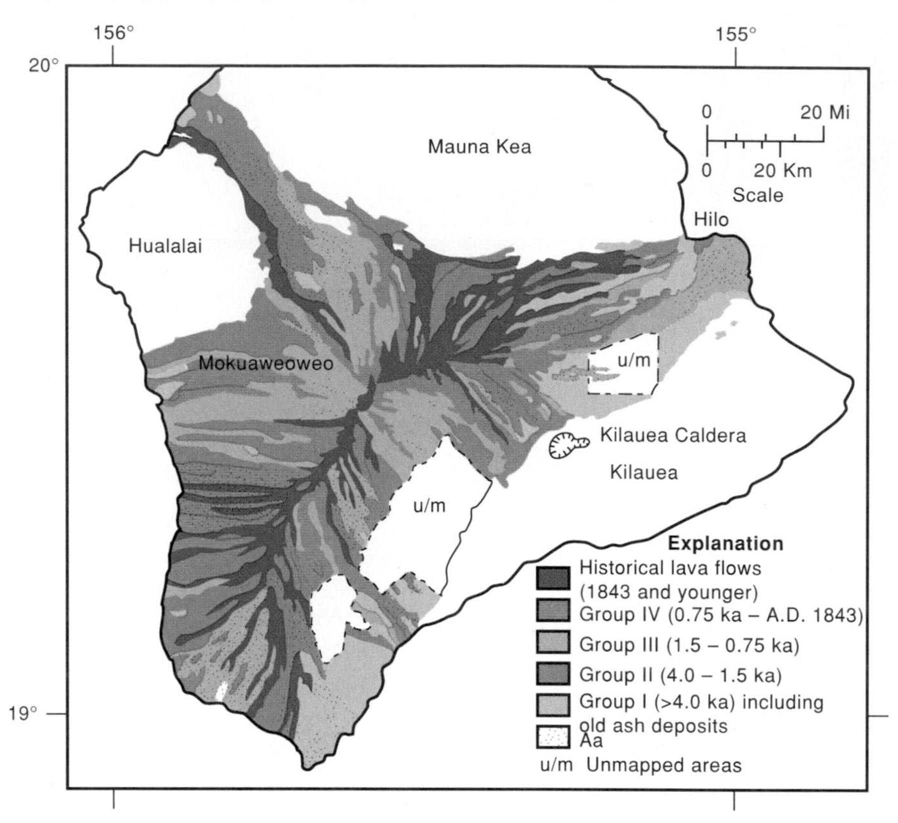


breathing tephra-laden air; damage to property is caused by the weight of tephra and its smothering and abrasive effects.

Glowing avalanches (nuées ardentes, pyroclastic flows, glowing clouds) are masses of incandescent, dry rock debris that move downslope like a fluid. They owe their mobility to hot air and other gases mixed with the debris and can travel many miles at speeds up to 100 miles per hour down valley floors on the flanks of a volcano. The path of an avalanche is guided largely by topography, but its great speed can cause it to climb vertically as much as several hundred meters, until it encounters opposing hill slopes or bends of the valley wall. This great mobility results when fragments in the moving mass are separated from each other and from the ground below by a cushion of hot, expanding gas, which largely eliminates friction as the mass moves. These flows can affect areas 15 miles or more from a volcano. Most losses from a pyroclastic flow are caused by the swiftly moving basal flow of hot rock debris, which can bury and incinerate everything in its path, and by the accompanying cloud of hot dust and gases, which can cause asphyxiation and burn lungs and skin.

Another common result of explosive eruptions is *mudflows* (*lahars*), masses of water-saturated rock debris that move downslope like flowing wet concrete. The debris comes from fragments of rock on the volcano flanks, and the water can come from rain, melting glacial ice and snow, a crater lake, or a reservoir adjacent to the volcano. Mudflows can be either hot or cold, depending on whether they contain hot rock debris. The speed of mudflows depends on their fluidity and the slope of the terrain; they sometimes move 50 miles or more down valley floors at speeds exceeding 20 miles per hour. Volcanic mudflows can reach even greater distances—about 60 miles from the source—than do pyroclastic flows (figure 8.2). The chief threat to humans is burial. Structures can be buried or swept away by the vast carrying power of the mudflow.

TABLE 8.1

Human Fatalities From Volcanic Activity, 1600–1986

| Primary Cause of Fatalities | 1600–1899 | | 1900–1986 | |
|---|---|---|---|---|
| Pyroclastic flows and debris avalanches | 18,200 | (9.8%) | 36,800 | (48.4%) |
| Mudflows (lahars) and floods | 8,300 | (4.5%) | 28,400 | (37.4%) |
| Tephra falls and ballistic projectiles | 8,000 | (4.3%) | 3,000 | (4.0%) |
| Tsunami | 43,600 | (23.4%) | 400 | (0.5%) |
| Disease, starvation, etc. | 92,100 | (49.4%) | 3,200 | (4.2%) |
| Lava flows | 900 | (0.5%) | 100 | (0.1%) |
| Gases and acid rain | ... | ... | 1,900 | (2.5%) |
| Other or unknown | 15,100 | (8.1%) | 2,200 | (2.5%) |
| Total | 186,200 | (100%) | 76,000 | (100%) |
| Fatalities per year (average) | 620 | | 880 | |

Values in parentheses refer to percentages relative to total fatalities for each time period.

*From R. I. Tilling, "Volcanic Hazards and Their Mitigation: Progress and Problems" in* Reviews of Geophysics, *27: 237–269, 1989, © Copyright by the American Geophysical Union.*

Lavas are generally erupted quietly, but can be preceded by explosive volcanic activity. The fronts of lava flows usually advance at less than human walking speed and hence cause no direct danger to human life (table 8.1). Generally, however, they totally destroy the areas they cover. Lava flows that extend into areas of snow can melt it and cause floods and mudflows; lava flows that extend into vegetated areas can start fires. The diameter of the hazard zone from lava flows is normally only a few miles because the lava solidifies as it moves; destruction is total but the affected area is relatively small.

## *Magnitude of Volcanic Eruptions*

The magnitude of volcanic eruptions is classified by the total amount of ejected material, both lava and pyroclastic debris (table 8.2). One fact evident in the table is that nature's scales are vastly different from the human scale. The disastrous explosive activity of Mount St. Helens in 1980, for example, is of the highest magnitude on the human scale, but only a minor feature on nature's scale.

## *Hazard Identification, Assessment, and Zonation*

Hazard assessments exemplify a familiar geologic adage: the present is the key to the past. To make such assessments, we must assume that a volcano will probably experience the same kinds of eruptive events, in the same general areas and at about the same average frequency, in the future as it has in the past (tables 8.3, 8.4). The danger inherent in this assumption has surfaced frequently, sometimes with disastrous results, but we have no practical alternative.

Figure 8.2

Sketch map showing the extent of two mudflows on opposite sides of Mount Ranier, Washington. The flows are confined to existing river valleys until the valleys become shallow enough for the mudflows to overflow the banks, about 25 miles from the volcano. *Source: Hays, 1981, p. 95, U.S. Geological Survey* Prof. Paper 1350.

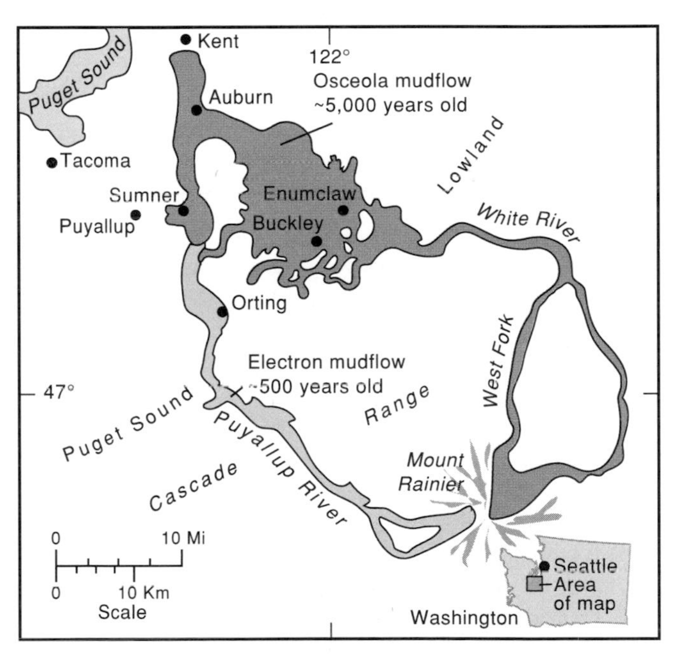

## TABLE 8.2a

### Magnitude of Volcanic Eruptions

| Magnitude | Volume of Ejected Material ($m^3$) | Magnitude | Volume of Ejected Material ($m^3$) |
|---|---|---|---|
| 1 | $10^9$ (1 billion) | 5 | $10^6$–$10^5$ |
| 2 | $10^9$–$10^8$ | 6 | $10^5$–$10^4$ |
| 3 | $10^8$–$10^7$ | 7 | $10^4$–$10^3$ |
| 4 | $10^7$–$10^6$ | 8 | Less than 1,000 |

## TABLE 8.2b

### Some Famous Eruptions

| Volcano | Volume New Material (billions of $m^3$) |
|---|---|
| Crater Lake (Mt. Mazama) | 42 |
| Tambora (1915) | 25 |
| Krakatoa (1883) | 18 |
| Mt. St. Helens (3,000 years ago) | 8.0 |
| Mt. St. Helens (450 years ago) | 2.6 |
| Vesuvius (79 A.D.) | 2.6 |
| Mt. St. Helens (1980) | 1.6–2.0 |
| Mt. Lassen (1914) | 1.0 |

## TABLE 8.3

### Proposed Criteria for Identification of High-Risk Volcanoes

| Hazard Rating | Score |
|---|---|
| 1. High silica content of eruptive products (andesite /dacite /rhyolite) | |
| 2. Major explosive activity within last 500 yr | |
| 3. Major explosive activity within last 5,000 yr | |
| 4. Pyroclastic flows within last 500 yr | |
| 5. Mudflows within last 500 yr | |
| 6. Destructive tsunami within last 500 yr | |
| 7. Area of destruction within last 5,000 yr is > 10 $km^2$ | |
| 8. Area of destruction within last 5,000 yr is > 100 $km^2$ | |
| 9. Occurrence of frequent volcano-seismic swarms | |
| 10. Occurrence of significant ground deformation within last 50 yr | |
| **Risk Rating** | |
| 1. Population at risk > 100 | |
| 2. Population at risk > 1,000 | |
| 3. Population at risk > 10,000 | |
| 4. Population at risk > 100,000 | |
| 5. Population at risk > 1 million | |
| 6. Historical fatalities | |
| 7. Evacuation as a result of historic eruption(s) | |
| | **Total Score** _____ |

A score of 1 is assigned for each rating criterion that applies; 0 if the criterion does not apply.

*Yokoyama, I., Tilling, R. I., and Scarpa, R. (1984). International Mobile Early-Warning System(s) for Volcanic Eruptions and Related Seismic Activities. Report of a UNESCO-UNEP-sponsored preparatory study in 1982–84. UNESCO, Paris.*

Figure 8.3a

The hazards zonation map for Nevada del Ruiz Volcano, Colombia. Although this map accurately anticipated the nature and areal extent of potential volcanic hazards and was available more than a month before the catastrophic eruption on November 13, 1985, its usefulness was negated by ineffective emergency management during the disaster.

*Source: Data from Herd and the Comite de Estudios Vulcanologics, 1986, figure 4.*

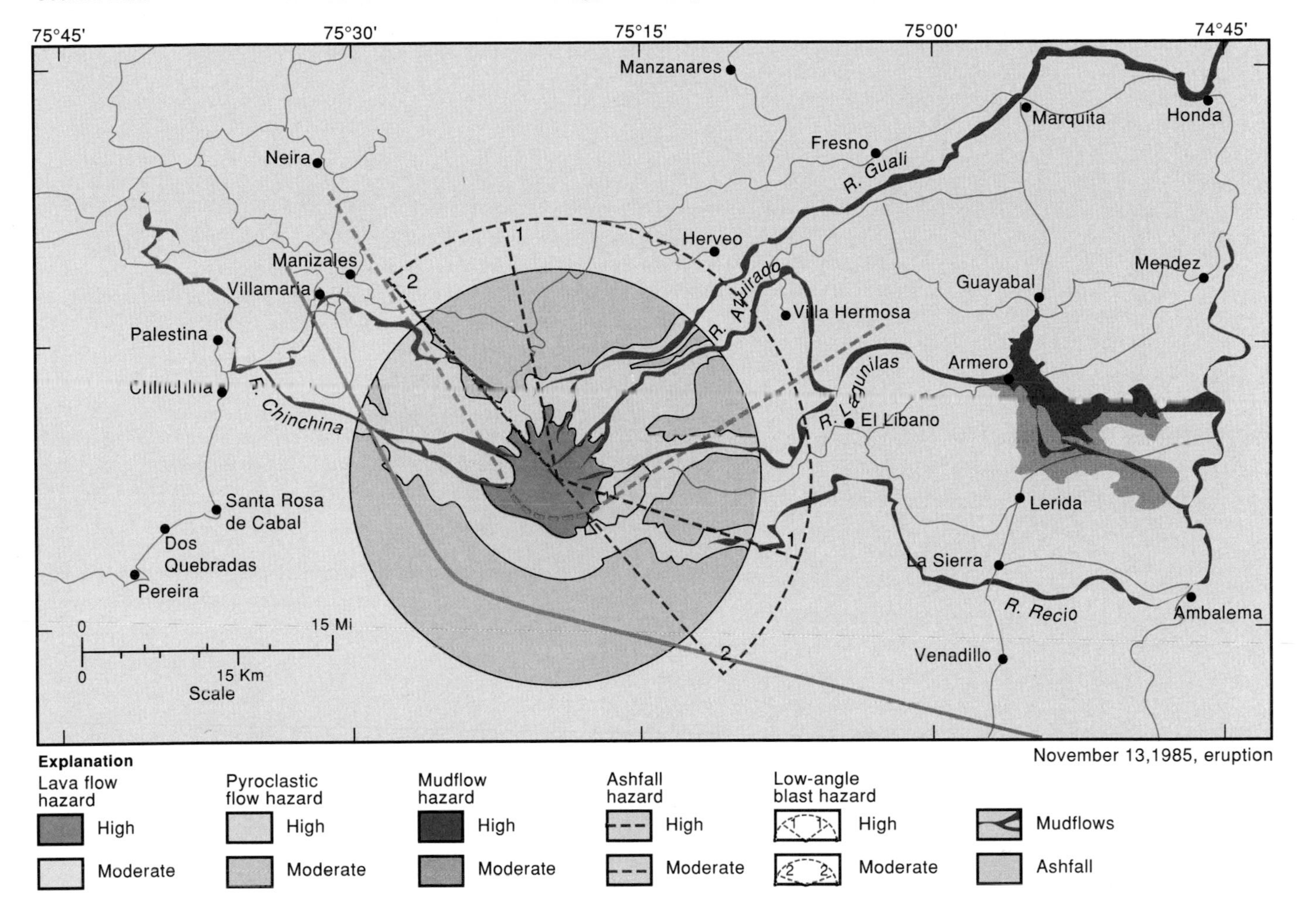


Figure 8.3 shows hazard zonation maps prepared for an area in Columbia and for the Big Island in Hawaii. Unfortunately, such maps are not yet available for many potentially dangerous volcanoes.

The incompleteness of available data poses a serious problem for public awareness of imminent danger from volcanic activity. The best that scientists can do, even with good data, is to estimate the likelihood of an eruption. Many times scientists' estimates prove incorrect, causing public skepticism regarding future pronouncements. Lawsuits might be lodged by those who either disrupted their lives for an eruption that failed to occur or lost relatives and/or property because a prediction was too imprecise. Scientists must continue their efforts to inform the public about the uncertainties inherent in predicting the future. The public tends to perceive scientists as Einstein-like figures who do not make errors. After all, if we can send someone to the moon, we should be able to do something as minor as recognizing an oncoming volcanic eruption—particularly from a volcano that has erupted before!

Land use around a suspect but currently dormant volcano poses a related problem. Should the government do more than simply inform the public about possible dangers? Should it prohibit someone from building a house on the flank of a certain volcano? Who is to decide what constitutes an acceptable level of risk? Obviously, these questions are political rather than scientific and must be decided by the residents of individual communities or states.

## TABLE 8.4

### Description of Volcanic Hazards

| | Lava Flows |
|---|---|
| Origin and characteristics | Result from nonexplosive eruptions of molten lava.<br>Flows are erupted slowly and move relatively slowly; usually no faster than a person can walk. |
| Location | Flows are restricted to areas downslope from vents; most reach distances of less than 6 miles. Distribution is controlled by topography.<br>Flows occur repeatedly at central-vent volcanoes, but successive eruptions may affect different flanks. Elsewhere, flows occur at widely scattered sites, mostly within volcanic "fields." |
| Size of area affected by single event. | Most lava flows cover no more than a few square miles. Relatively large and rare flows probably would cover only hundreds of square miles. |
| Effects | Land and objects in affected areas subject to burial, and generally they cause total destruction of areas they cover. Those that extend into areas of snow may melt it and cause potentially dangerous and destructive floods and mudflows.<br>May start fires. |
| Predictability of location of area endangered by future eruptions. | Relatively predictable near large, central-vent volcanoes. Elsewhere, only general locations predictable. |
| Frequency, in conterminous United States as a whole. | Probably one to several small flows per century that individually cover less than 10 square miles. Flows that cover tens to hundreds of square miles probably occur at an average rate of about one every 1,000 years. (In Hawaii, eruption of many flows per decade would be expected.) |
| Degree of risk in affected area. | To people, low.<br>To property, high. |

*Mullineaux, in Hays, U.S. Geological Survey, 1981, p. 88.*

| Hot Avalanches, Mudflows, and Floods | Volcanic Ash (Tephra) and Gases |
| --- | --- |
| Hot avalanches can be caused directly by eruption of fragments of molten or hot solid rock; mudflows and floods commonly result from eruption of hot material onto snow and ice and eruptive displacement of crater lakes. Mudflows also commonly caused by avalanches of unstable rock from volcano.<br>Hot avalanches and mudflows commonly occur suddenly and move rapidly, at tens of miles per hour. | Produced by explosion or high-speed expulsion of vertical to low-angle columns or lateral blasts of fragments and gas into the air; materials can then be carried great distances by wind. Gases alone may issue nonexplosively from vents.<br>Commonly produced suddenly and move away from vents at speeds of tens of miles per hour. |
| Distribution nearly completely controlled by topography.<br>Beyond volcano flanks, effects of these events are confined mostly to floors of valleys and basins that head on volcanoes.<br>Large snow-covered volcanoes and those that erupt explosively are principal sources of these hazards. | Distribution controlled by wind directions and speeds, and all areas toward which wind blows from potentially active volcanoes are susceptible. Zones around volcanoes are defined in terms of whether they have been repeatedly and explosively active in the last 10,000 years. |
| Deposits generally cover a few square miles to a few hundreds of square miles. Mudflows and floods may extend downvalley from volcanoes many tens of miles. | An eruption of "very large" volume could affect tens of thousands of square miles, spread over several States. Even an eruption of "moderate" volume could significantly affect thousands of square miles. |
| Land and objects subject to burning, burial, dislodgement, impact damage, and inundation by water. | Land and objects near an erupting vent subject to blast effects, burial, and infiltration by abrasive rock particles, accompanied by corrosive gases, into structures and equipment. Blanketing and infiltration effects can reach hundreds of miles downwind. Odor, "haze," and acid effects may reach even farther. |
| Relatively predictable, because most originate at central-vent volcanoes and are restricted to flanks of volcanoes and valleys leading from them. | Moderately predictable. Voluminous ash originates mostly at central-vent volcanoes; its distribution depends mainly on winds. Can be carried in any direction; probability of dispersal in various directions can be judged from wind records. |
| Probably one to several events per century caused directly by eruptions.<br>Probably only about one event per 1,000 years caused directly by eruption at "relatively inactive" volcanoes. | Probably one to a few eruptions of "small" volume every 100 years. Eruption of "large" volume may occur about once every 1,000 to 5,000 years. Eruption of "very large" volume, probably no more than once every 10,000 years. |
| Moderate to high for both people and property near erupting volcano. Risk relatively high to people because of possible sudden origin and high speeds. Risk decreases gradually downvalley and more abruptly with increasing height above valley floor. | Moderate risk to both people and property near erupting volcano; decreases gradually downwind to very low. |

**Figure 8.3b**

Map of the Big Island showing the volcanic hazards from lava flows. Severity of the hazard increases from zone 9 to zone 1. Shaded areas show land covered by historic flows from three of Hawaii's five volcanoes (Hualalai, Mauna Loa, and Kilauea).

*Source: Tilling et al., 1987, p. 49, U.S. Geological Survey.*

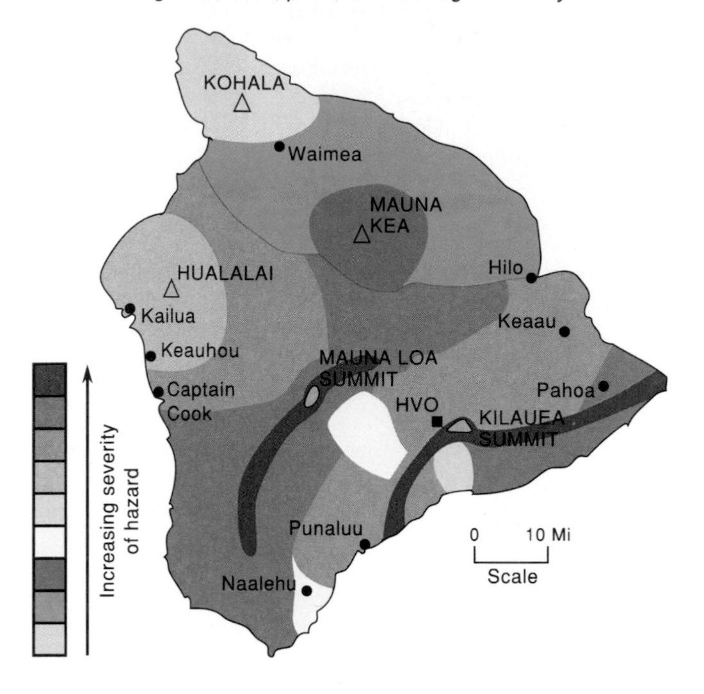


## *Volcanoes and Climate*

In addition to its immediate danger to nearby residents, an erupting volcano poses a more far-reaching and long-lasting danger. Major eruptions hurl both fine-grained volcanic dust and sulfur dioxide gas into the atmosphere (figure 8.4), where water and oxygen rapidly convert the sulfur dioxide into sulfuric acid. In addition to forming acid rain, the sulfuric acid blocks solar radiation, resulting in lower temperatures at the Earth's surface. The April, 1815, eruption of Tambora volcano in Indonesia, for example, spewed out 150 $km^3$ of ash and pumice, cut sunlight by 25%, caused the coldest summer in New Haven, Connecticut, in 200 years, and caused many crop failures in the Northern Hemisphere. In central England the summer of 1816 was about 1.5°C cooler than the previous summer. This dismal weather is credited with inspiring Mary Shelley to write *Frankenstein,* and Lord Byron his poem "Darkness." A very large eruption in the near future might drastically affect crop yields and create and exacerbate food shortages in many areas, especially in the marginally productive regions where some of the world's poorest people live. Eruptions such as that of Tambora constitute a very real volcanic hazard in terms of the number of people affected. There is no question that such large eruptions will recur; the only uncertainty lies in where and when.

**Figure 8.4a**

Geologist sampling gases and measuring temperatures at Deformes fumarole (gaseous volcanic vent) on the rim of the inner crater of Galeras Volcano, Colombia. The temperature is about 500 ° F, and the gases contain up to 5,000 tons/day of $SO_2$, which is transformed to sulfuric acid in the atmosphere. The yellow material around the rim of the crater is sulfur.

## Problems

1. Table 8.1 shows that the average number of fatalities per year from volcanic activity was higher in the 87-year period between 1900 and 1986 than in the previous 300 years. Explain why.
2. Examine a geologic map of the United States to answer the following questions:
   a. Where has volcanic activity occurred during the past 2 million years (Quaternary Period)?
   b. Where has volcanic activity occurred during the past 65 million years (Cenozoic Era)? Is this area larger or smaller than the area of Quaternary activity? Explain why.
   c. How does the area covered by pre-Cenozoic volcanic material compare in size with the area covered by Cenozoic activity? Explain the reason for the difference.
3. Compare figures 8.1 and 8.3B, which show the Big Island (Hilo), Hawaii. Does the temporal sequence of lava flows in figure 8.1 match the hazard zonation map of figure 8.3B? What factors other than the temporal sequence of lava flows might contribute to preparation of the hazard zonation map?
4. How might you explain a difference among eruptions in the size and areal distribution of fragments ejected from a volcano?
5. In humid climates soils form very rapidly on basaltic tephra and flow rocks after they cool. These soils are also unusually fertile. Explain these observations.
6. Basaltic lavas tend to flow quietly, whereas rhyolitic lavas tend to come to the Earth's surface explosively. What might explain these tendencies? How might this

**Figure 8.4b**

Dispersal of $SO_2$ gas with time from the eruption of Mt. Pinatubo in the Phillippines in 1991.

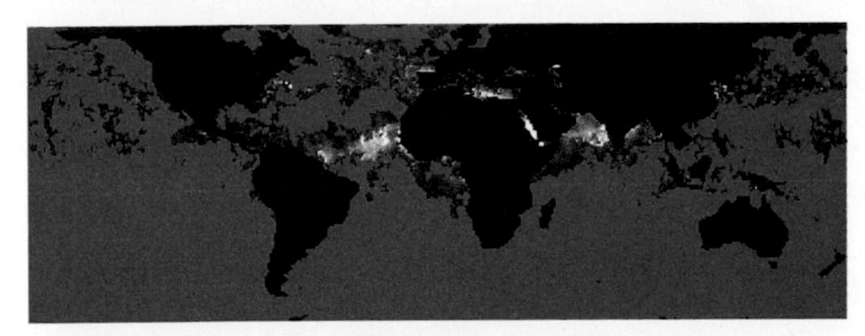

Week 1

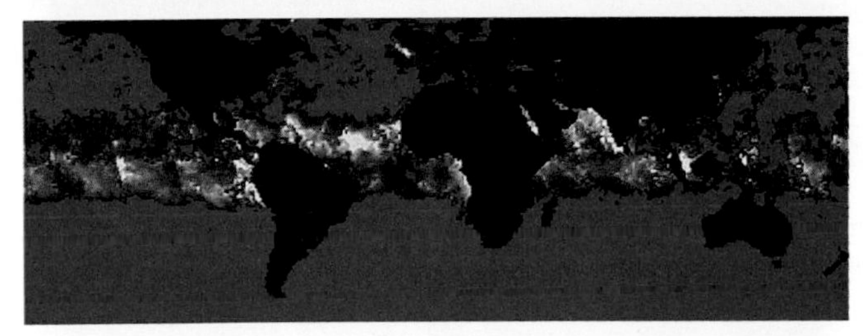

Week 6

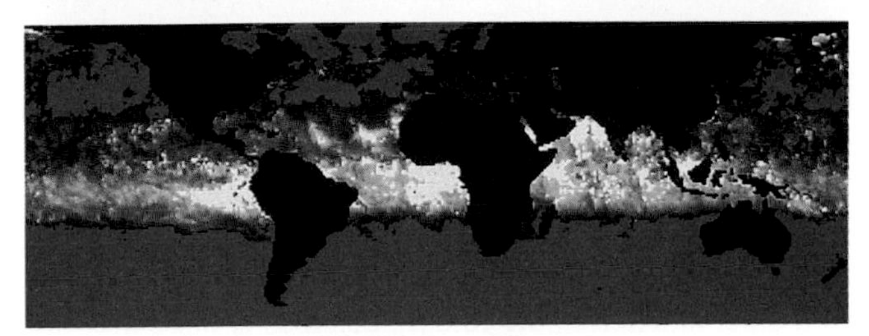

Week 12

difference in eruptive style be reflected in the shape of the volcanic accumulations formed from basaltic and rhyolitic lavas?

7. The 1982 eruption of the Mexican volcano El Chichón caused the emission of very large amounts of sulfur-rich gas into the atmosphere. The gas produced clouds of sulfuric acid droplets that spread around the Earth. What effects do you think this sulfuric acid might have had on the Earth's surface as the droplets fell to the ground?
8. You have gambled and apparently lost. You built your home adjacent to a dormant volcano, and now the volcano is starting to erupt, spilling lava from the lava lake in its crater. The lava is heading your way and you have let your insurance lapse because the annual premium increased so much. You estimate it will take about two days before you are ruined financially—unless you can think of possible ways to stop or divert the lava before it enters your front door. Can you think of any schemes that might save your family farm?
9. Too little money is available to permit permanent geological and geophysical monitoring of all volcanoes. What priority should expanding this program have in relation to other social needs?
10. Different scientists typically have different views about the likelihood that an event of some kind will occur. Such disagreement confuses the public. How can scientists avoid confusing the public when the issue is the likelihood of a volcanic eruption?
11. Whose responsibility should it be to protect people against the risk of living within "striking distance" of a volcano that shows some of the signs that commonly precede an eruption?

## Further Reading/References


Bolt, B. A., Horn, W. L., MacDonald, G. A., and Scott, R. F., 1975, *Geological Hazards*. New York, Springer-Verlag, 328 pp.

Hays, W. W. (ed.), 1981, *Facing Geologic and Hydrologic Hazards*. U.S. Geological Survey Professional Paper 1240-B, 108 pp.

Latter, J. H. (ed.), 1989, *Volcanic Hazards. Assessment and Monitoring*. New York, Springer-Verlag, 625 pp.

Martinelli, B., 1991, Understanding triggering mechanisms of volcanoes for hazard evaluation: *Episodes*, v. 14, pp. 19–25.

Peterson, D. W., 1988, Volcanic hazards and public response: *Journal of Geophysical Research*, v. 93, no. B5, pp. 4161–4170.

Rampino, M. R., Self, S., and Stothers, R. B., 1988, Volcanic winters: *Annual Review of Earth and Planetary Sciences*, v. 16, pp. 73–99.

Scott, W. E., 1989, Volcanic-hazard zonation and long-term forecasts: in *Short Course in Geology*, v. 1, *Volcanic Hazards*, R. I. Tilling (ed.), pp. 25–49, Washington, D.C., American Geophysical Union.

Tazieff, H., and Sabroux, J. C. (eds.), 1983, *Forecasting Volcanic Events*. New York, Elsevier, 635 pp.

Tilling, R. I., 1989, Volcanic hazards and their mitigation: Progress and problems: *Reviews of Geophysics*, v. 27, pp. 237–269.

Tilling, R. I., Heliker, C., and Wright, T. L., 1987, *Eruptions of Hawaiian Volcanoes*. U.S. Geological Survey, 54 pp.

Van Rose, S., and Mercer, I. F., 1991, *Volcanoes*, 2nd ed., British Museum, 60 pp.

Wood, C. A., and Kienle, J., 1990, *Volcanoes of North America*. New York, Cambridge University Press, 354 pp.

Wright, T. L., and Pierson, T. C., 1992, *Living with Volcanoes*. U.S. Geological Survey Circular 1073, 57 pp.

exercise

# NINE

## Fluvial Processes and Forms

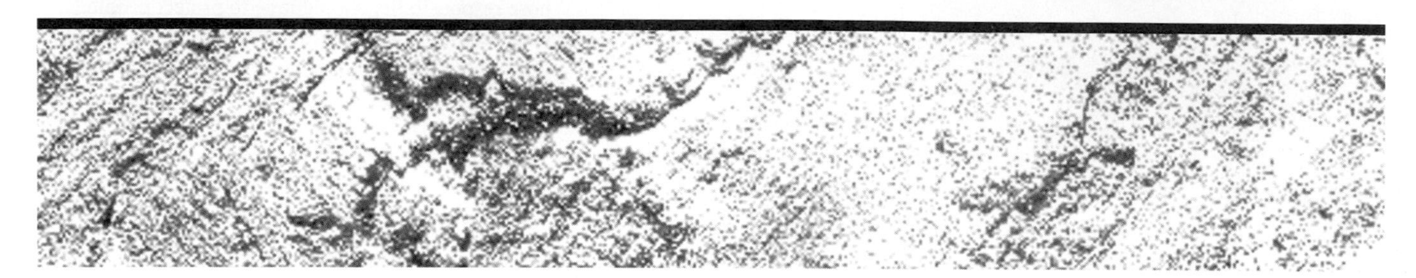

Streams are one of the most important features on the Earth's surface. Stream courses that are dry throughout most of the year, bearing water only during and immediately after a rain, are called *ephemeral*. Stream channels that carry water during one part of the year, are dry during the other, and are fed by underground water are called *intermittent*. Streams that carry water continuously and are fed both by overland flow and from below are called *perennial*. The more humid the climate, the higher the proportion of perennial streams.

Total runoff on the ground surface depends on rainfall, evaporation, transpiration by plants, and infiltration into the soil. In general, the higher the amount of rainfall, the higher the runoff. For many communities, the runoff that supplies lakes and rivers controls both the quantity and the quality of the local water supply. If local rivers or lakes have a high enough capacity to supply the needed volume of water all year, a community can draw from this source continuously. If water requirements exceed minimum streamflow, the community can store water until needed, although ponding increases the surface area of the water and therefore the amount of evaporation, reducing the ultimate yield. If its surface water supplies are inadequate, the community must tap underground sources. Water supply places the ultimate limitation on the number of people who can live in an area.

The catchment area of a stream is called its watershed or *drainage basin* and includes the entire area the stream serves. The famous Continental Divide in the western United States is an imaginary line that separates water channels that drain into the Pacific Ocean from those that drain into the Atlantic. Successively smaller drainage basins on the Atlantic Ocean side include the Gulf of Mexico basin, the Mississippi River basin, the Arkansas River basin, the Canadian River basin, and so on, down to the small drainage basin of the creek that flows by your house. Both the size and the shape of a drainage basin are important in determining runoff. Basin size, of course, has a strong effect on total runoff; the larger the catchment area, the greater the runoff. Basin shape is also important, however, because it influences the temporal distribution of runoff. In an elongate basin, flow in tributary channels reaches the main stream at different times, distributing the runoff over a long time span. In a more equant drainage basin, tributaries feed into the main stream at about the same time, resulting in a sudden high peak flow.

Topography also affects runoff, which decreases on gentler slopes. Gentle slopes allow more time for infiltration and for at least temporary water storage. They also tend to be more densely vegetated than extremely steep slopes, and vegetation decreases runoff because plant roots hold water. Soil and surface-sediment character can be an important factor as well; loose, permeable sediment permits easy infiltration of water and thus decreases runoff.

### *Hydrographs*

A *hydrograph* shows how streamflow varies with time and, therefore, reflects rainfall duration and intensity as well as the characteristics of the drainage basin that influence runoff. Hydrographs are plots of water discharge versus

**Figure 9.1**

Long-period hydrograph for one point on Horse Creek near Sugar City, Colorado, showing the stream discharge over a period of one year.

*Source: U.S. Geological Survey* Open-File Report 79-681.

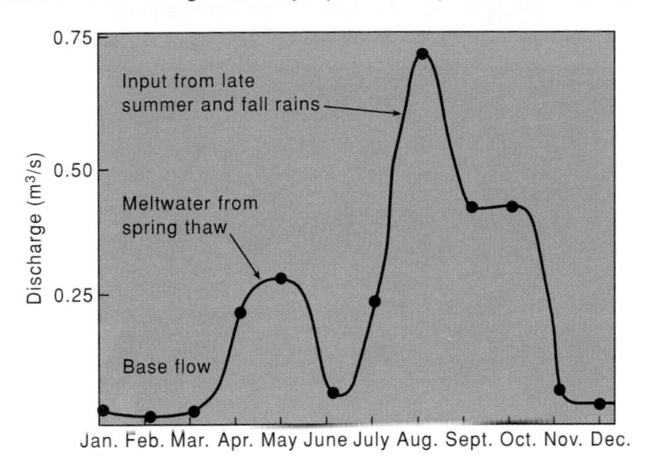


**Figure 9.2**

Short-period flood hydrographs for two different points along Calaveras Creek near Elmendorf, Texas. The flood has been caused by heavy rainfall in the drainage basin. In the upstream part of the basin, flooding quickly follows the rain. The larger stream lower in the drainage basin responds more sluggishly to the input.

*Source: U.S. Geological Survey Water Resources Division.*

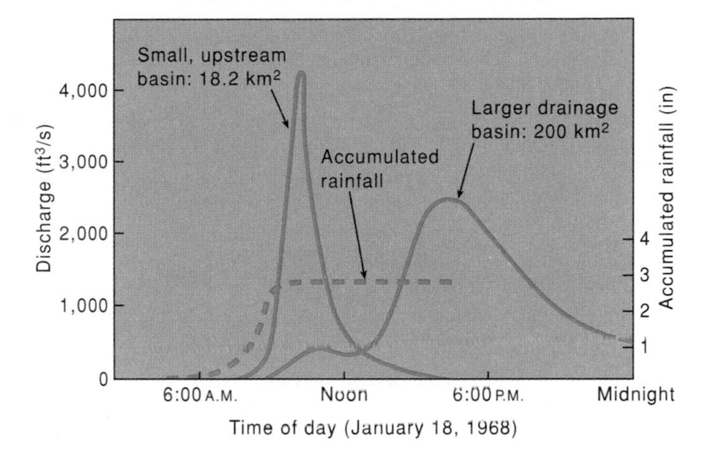


time (figure 9.1) that permit environmental geologists to determine the total flow, the base flow (upward flow from the groundwater into the stream), and periods of high (flood) and low flows. Long-period hydrographs (with graph axes calibrated in months or years) are used in designing irrigation projects and dam construction and for forecasting floods. Short-period hydrographs (with axes calibrated in hours or days) are used to show peak discharges during floods (figure 9.2). Stream discharges are calculated from the equation

$Q = AV$, where

- $Q$ = discharge (ft³/sec)
- $A$ = cross-sectional area of the stream channel at the location where discharge is being determined (stream width multiplied by stream depth; ft²)
- $V$ = mean velocity at the site (ft/sec)

## *Effects of Water Movement*

The movement of channelized water has three consequences important to environmental geologists: erosion, sediment transport, and sediment deposition. Moving water has the ability to transport sediment; the amount transported depends on the amount of water, its velocity and gradient, and other factors. The moving sediment causes downcutting of the stream channel by abrasion, as well as headward erosion of the stream. The channel is thus deepened and lengthened, enlarging its drainage basin and increasing topographic relief. Downcutting also widens the channel by undermining its sides, causing them to collapse into the moving water and be transported downstream. The erosion process at first increases local relief, and then decreases it as the stream valley widens. Neighboring streams can be at very different stages of this process, depending on factors such as local variation in types of bedrock or loose sediment, rainfall intensity, or human intervention in the form of dam construction, plowing, or spreading of concrete for highways and buildings.

Streams transport material in three ways: by traction (bed load), by suspension, and by solution. In traction transport sediment either rolls along the stream bottom or moves in a hopping fashion, called *saltation.* The moving water creates stress on the stream bottom because of friction between the water and the individual grains. This friction generates upward eddies of water strong enough to lift the grains off the bottom so they can be moved downstream. The higher the degree of stress, the larger the grain size and the greater the number of grains the stream can move. The size of the largest particle the stream can move is termed the *competence* of the stream. Competence varies as the sixth power of stream velocity. The velocity of a stream is determined mostly by its *gradient,* expressed as the number of feet (or meters) the stream descends for each mile (or kilometer) along its flow path. Gradients are steepest in headwater tributaries; in mountainous areas they can exceed 50 m/km. The lower reaches of the Mississippi River, in contrast, have gradients of only 1 or 2 cm/km.

*Capacity* is the total amount of sediment a stream can carry. Most streams with high capacities also have high suspension loads. Usually the capacity of a stream varies as the third power of the velocity.

Muddy sediment erodes less easily than sandy sediment because it is more cohesive. Clay flakes are shaped like pieces of paper and adhere to each other almost immediately when they make contact at the stream bottom; the adhesion makes them more difficult for the moving water to erode, or pick up. Once they are picked up, however, they are transported easily because of their small size, so a stream can transport large volumes of mud very rapidly. The amount of suspended load in a stream depends mostly on water discharge rather than on bottom shear stress.

Solution load is important in water-quality investigations. In unpolluted areas the ions in the water have dissolved from the rocks in the drainage basin. Waters affected by human activities, in contrast, can contain almost anything.

As a stream progresses through its lifetime of sculpting the land surface, it creates features that are very important in environmental geology. Most of these features develop during the middle part of the stream's life cycle, as the stream changes its major work from downcutting (vertical erosion) to lateral cutting (horizontal erosion). As lateral erosion proceeds, the stream channel begins to meander (figure 9.3). With time, the meanders enlarge and move downstream; their rate of movement depends on water discharge and local geology. Meanders can move laterally at speeds of tens or even hundreds of meters per year, although rates below 10 m/yr are more common on smaller streams.

Meanders develop best in muddy streams—streams with a high ratio of suspension load to bed load. The meanders migrate because the deepest, swiftest, and most turbulent section of the stream channel,where erosion is most active, lies along the outer margin of each meander bend. Slower, less turbulent water along the inside of the bend results in sediment deposition. Meanders thus become increasingly exaggerated, and the width of the meander belt increases. As the width increases, meanders are more easily cut off from the stream to form *oxbow lakes.* These lakes often have swampy, poorly drained areas rich in organic matter.

When a stream overflows its channel, or floods, it creates raised ridges called *levees* at the channel margins. Levees develop because water velocity decreases sharply as it leaves the channel, causing suspended sediment to be

**Figure 9.3**

The evolution of stream meanders results from both erosion on the outside of a curve in the stream channel, where velocity is greatest, and deposition on the inside of the curve, where velocity is lowest. (a) Streamflow is deflected by an irregularity and moves to the opposite bank, where erosion begins. (b) Once the bend begins to form, the flow of water continues to impinge on the outside curve, so a meander loop develops. At the same time, deposition occurs on the inside of the bend as a result of the lower stream velocities in that area. (c) The meander is enlarged and migrates laterally, with the contemporaneous growth of a point bar. A general downslope migration of meanders occurs as they grow larger and ultimately cut themselves off to form oxbow lakes.



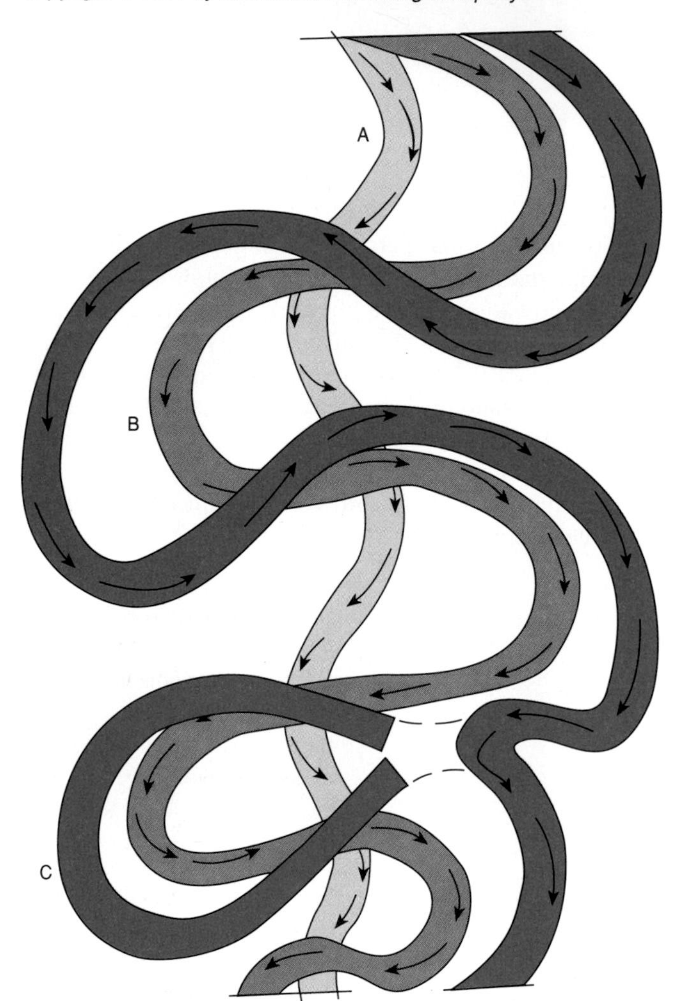

## Figure 9.4

The major features of a floodplain include meanders, point bars, oxbow lakes, natural levees, backswamps, and streams. A stream flowing around a meander bend erodes the outside curve and deposits sediment on the inside curve to form a point bar. The meander bend migrates laterally and is ultimately cut off, to form an oxbow lake. Natural levees build up the banks of the stream, and backswamps develop on the lower surfaces of the floodplain. Yazoo streams have difficulty entering the main stream because of the high natural levees, and thus flow parallel to it for considerable distances before becoming tributaries. Slope retreat continues to widen the low valley, which is partly filled with river sediment.

*Reprinted with the permission of Macmillan Publishing Company from* The Earth's Dynamic Systems, *Fifth Edition by W. Kenneth Hamblin. Copyright © 1989 by Macmillan Publishing Company.*

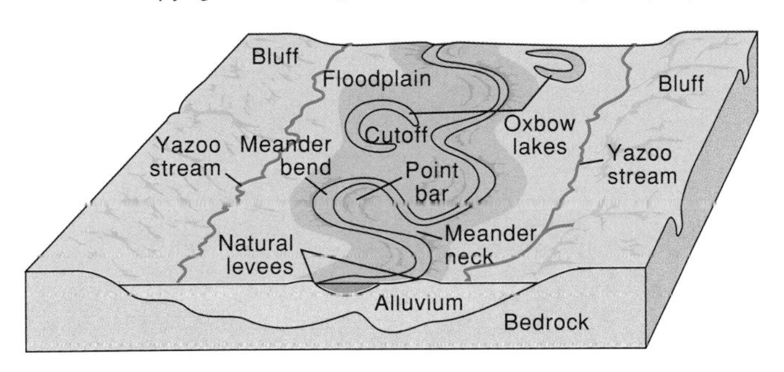


deposited almost immediately, and building a ridge at the channel margin. With each succeeding flood, the height of the levee increases.

The area into which the stream spills over during floods is called its *floodplain* (figure 9.4); this is the surface on which the meanders are located. The width of the meander belt can be no greater than the width of the floodplain, and is often much smaller. Some wide-floored valleys have elevated, nearly level benches called *terraces* along their margins (figure 9.5). Typically these benches occur at the same elevation on both sides of the channel. Terraces are remnants of former valley floors, valley-wide floodplains that once existed at higher levels than that of the present floodplain. The renewed downcutting that left the terrace might have been caused by either climatic change or tectonic uplift in the stream's headwaters area.

## Figure 9.5

Sketch of a valley filled with alluvium that has been eroded into terraces. Each terrace is coded by a pattern.

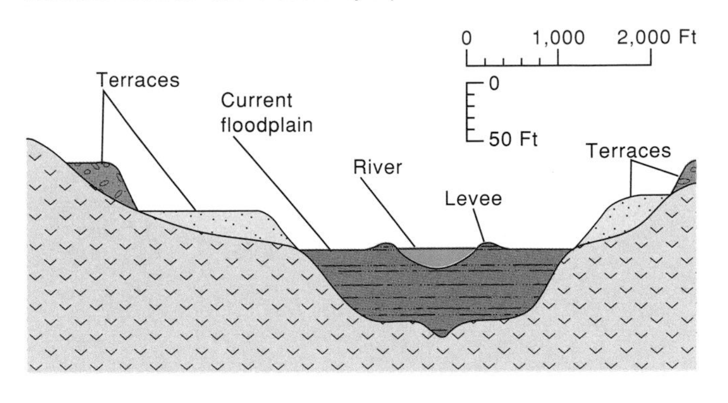


# Problems

1. Why is the volume of water in surface runoff always less than the volume of precipitation?
2. How do both the presence and the type of rock or sediment at the Earth's surface affect the amount of runoff? Compare, for example, the effects of granite versus gravel or sandstone versus shale.
3. Calculate the number of square feet of sediment removed in figure 9.5 between the higher terrace and the lower terrace; also between the lower terrace and the present floodplain. Is the total amount of sediment removed greater or less than the amount present in the streambed above the bedrock? How can you account for the difference?
4. In figure 9.4 the width of the meander belt is only about half the width of the stream valley. What can you infer from this?
5. In figure 9.6, assume that at point A in the river the elevation of the bottom is 70 feet, and at point B it is 65 feet. How much did the formation of the Caulk cutoff in 1937 change the gradient of the river?
6. What was the rate of movement of the meander bend from west to east between 1827 and 1846?
7. Why do the topographic contour lines in the Caulk Neck–Caulk Point region roughly parallel the meander?
8. What is the maximum local relief of the natural levees in this area?
9. The topographic map shows that the state boundary between Arkansas and Mississippi was drawn in the mid-1800s. By examining the map you can determine the approximate date when the state boundary was established between Arkansas and Mississippi. Estimate this date. What is your opinion of defining state boundaries that follow the course of a river?

**Figure 9.6**

Topographic map of a part of the Lamont, Arkansas Quadrangle.

CI 5 ft 0 1 mile

## Further Reading/References


Brookes, A., 1988, *Channelized Rivers.* New York, John Wiley & Sons, 326 pp.

Knighton, D., 1984, *Fluvial Forms and Processes.* London, Edward Arnold, 218 pp.

Morisawa, M., 1985, *Rivers.* New York, Longman.

Ritter, D. F., 1986, *Process Geomorphology,* 2nd ed. Dubuque, Iowa, Wm. C. Brown, 592 pp.

exercise

# TEN

## Floods

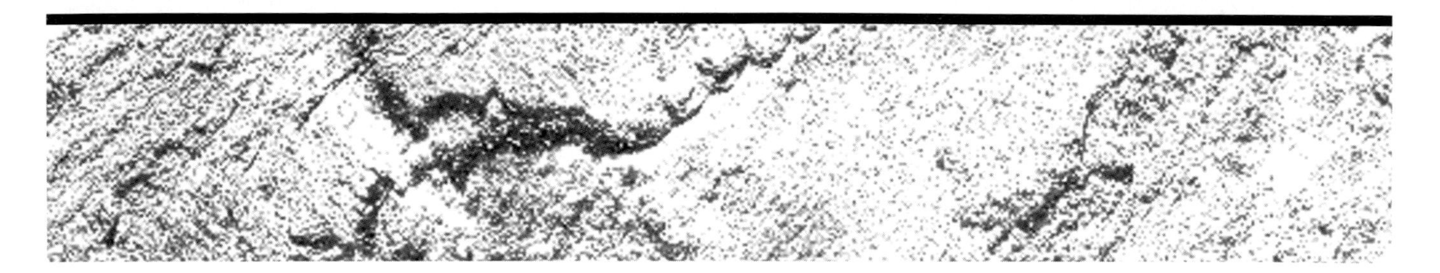

Floods are the most ubiquitous of the many geologic catastrophes that plague humankind; they affect more people than all other natural hazards combined. As much as 90% of the damage related to natural disasters (excluding droughts) is caused by floods, at an estimated annual cost of 2.4 billion dollars. About 7% of the land area of the 48 conterminous United States is subject to flooding, and these floods can cover hundreds of thousands of square miles (figure 10.1). More than 20,000 communities, with over 6 million single-family homes, and representing perhaps 10% of the total U.S. population, are located on flood-prone land. Of the recent major disasters declared by U.S. presidents, 85% were associated with floods.

Many floods have a common cause: too much rainfall within a short time. When rain falls slowly, enough water is absorbed by soil and bedrock or channelled into stream courses to prevent inundation of the surrounding area. If rainfall is excessive, however, flooding occurs, often with loss of life and extensive property damage (figure 10.2). Flooding can also occur if heavy snowfall melts quickly in abnormally warm spring weather. Other floods are secondary, resulting perhaps from breaching of a natural dam formed by a landslide across a stream, or from human disruption of the natural environment, as in the case of a dam failure. Flood control is a major environmental problem in many areas. In the United States an important area of concern is the lower Mississippi River. Overseas, the most continually endangered region might be the nation of Bangladesh, whose 75 million people live on 55,000 square miles of low-lying, perennially flood-prone land along the Ganges River.

The key controls over an area's tendency to flood are (1) the amount and distribution of precipitation and (2) the topographic and geologic characteristics of the rainfall catchment area. Average annual rainfall ranges from virtually zero in the driest deserts to 451 inches at one location in Hawaii. Perhaps more important than such averages, however, is the seasonal distribution of the precipitation. Clearly, if all the rain falls within a one-month period, flooding is more likely than if it falls fairly evenly throughout the year. Meteorologists understand the general controls that govern the temporal distribution of precipitation, but most flooding problems are caused by unusually heavy rainstorms that occur at irregular intervals.

Topography has a direct influence on flooding. The most areally extensive floods (though not necessarily the most damaging) occur in low-lying, downstream areas. These areas have larger catchments for collecting precipitation, but shallower stream channels for containing it. A *floodway* consists of the area of channel and immediately

## Figure 10.1a

Map showing distribution of great floods in the conterminous United States since 1889.
*Source: U.S. Geological Survey.*

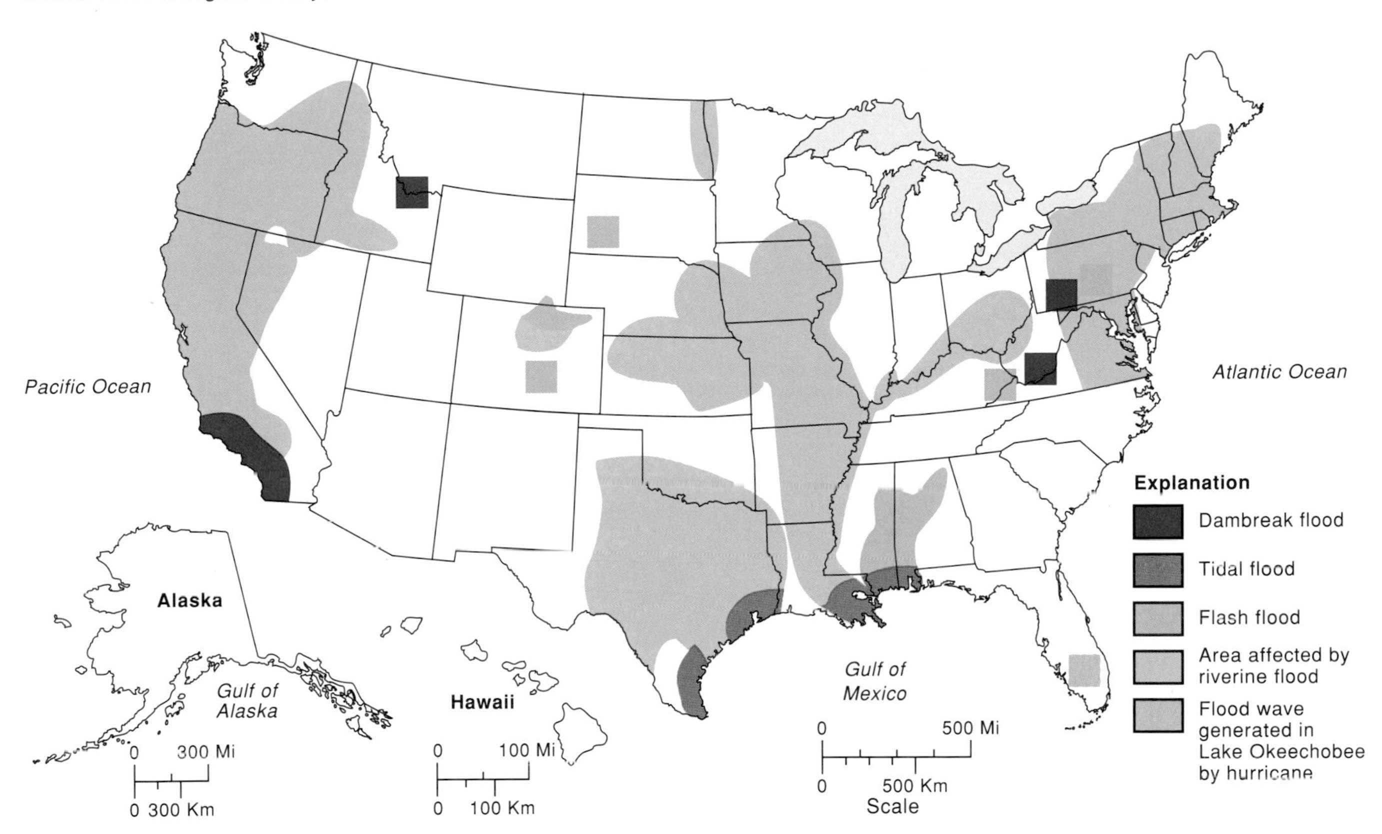


## Figure 10.1b

Curves showing the approximate limits of the largest floods experienced in the United States at successive times. The flattening of the curves at larger drainage areas indicates that the peak discharges per unit of drainage area are smaller as the size of the drainage basin increases. With the passage of time, the curves have moved up as larger floods have occurred. The longer the period for which records are available, the more likely it is that a flood will occur with an unusually large discharge.
*Source: U.S. Geological Survey.*

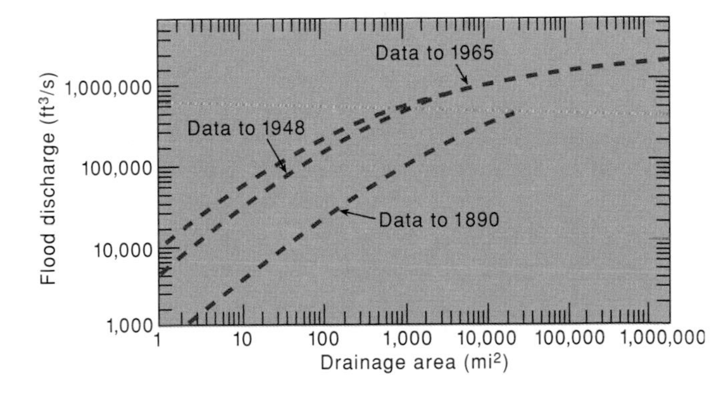


surrounding flat ground that provide the avenue for floodwaters (figure 10.3). In inhabited areas humans have important effects on the amount of runoff. By devegetating, bulldozing, building on, or otherwise changing the land surface, humans greatly affect the amount of water that infiltrates into the ground or runs off the surface. For example, clearing a forest increases the intensity with which rain hits the soil, reduces infiltration, and increases surface runoff into streams. Structures such as houses and pavement increase runoff by decreasing the natural surface available to soak up precipitation, and thereby increase the frequency of flooding. In 1974 the United States Geological Survey evaluated the extent and development of urban floodplains. Among the 26 moderate to large cities studied, an average of 52.8% of their total floodplain areas had been urbanized. Values such as 97% for Great Falls, Montana; 89.2% for Phoenix, Arizona; 83.9% for Tallahassee, Florida; and 83.5% for Harrisburg, Pennsylvania, indicate why so many urban areas are susceptible to yearly flooding and its associated damage.

## Figure 10.2

Flood frequency curve for Eel River in Scotia, California, based on data collected from 1932 to 1959. The graph shows how often on average a given discharge will occur.

*From Gary B. Griggs and John A. Gilchrist,* Geologic Hazards, Resources, and Environmental Planning, *2d ed. Copyright © 1983 Wadsworth Publishing Company, Belmont, CA. Reprinted by permission.*

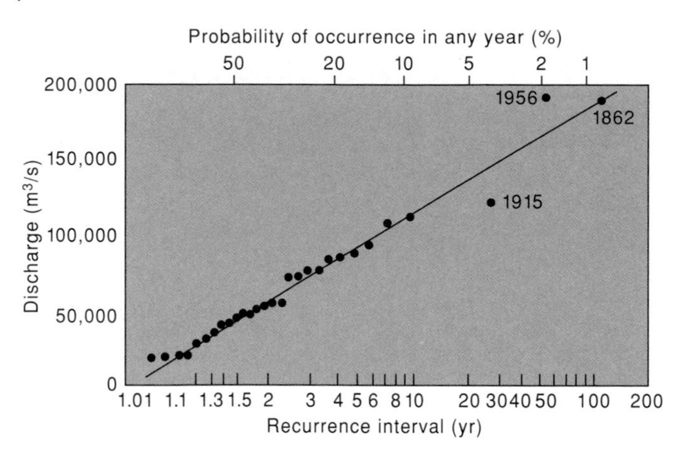


## Figure 10.3

Perspective sketch showing relationships of the river channel to bottomland. The cross-section shows flood stages and flood frequencies.

*From Moss and others, 1978.*

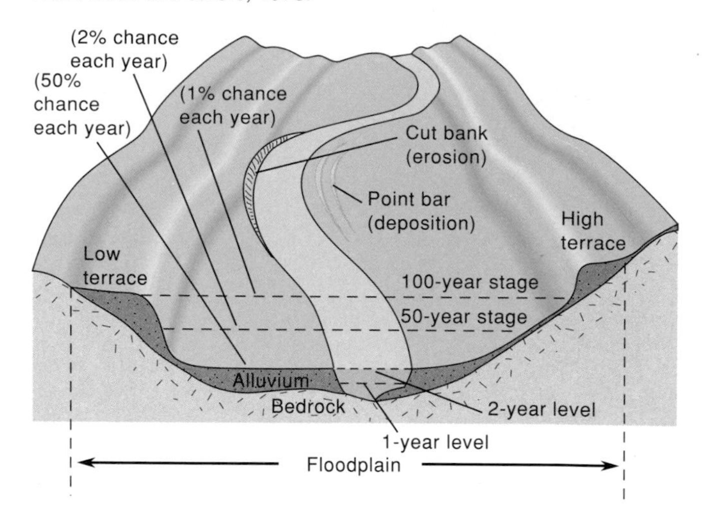


The amount and character of the debris carried in a flood is affected not only by the intensity of the precipitation but also by the nature of the bedrock. For example, a watershed underlain by poorly consolidated, fine-grained sedimentary rocks produces debris consisting of mud, while a watershed composed largely of harder rock yields coarser debris. Overland flows move these materials and transport them to stream channels. On steep slopes with loose, broken rock or poorly consolidated soils, downpours can trigger avalanches, slides, or mudflows. Structures located in the path of such a moving mass are often damaged by the impact. Solid materials derived from watershed slopes lodge in channels, later to be swept away by currents, usually during floods.

### *Flood Frequency Analysis*

The objective of a flood-frequency analysis is to determine how often, on the average, a particular region can expect a flood of a certain magnitude. The steps in the analysis are as follows:

1. Obtain the streamflow records of a particular gaging station for all the years during which records have been kept. Choose the station with the longest, most complete record.
2. Identify and list the highest discharge rate for each year.
3. Rank the water discharges in decreasing order.
4. To determine the recurrence interval for each discharge, use the formula

$$\text{Average Recurrence Interval} = \frac{n+1}{m}$$

where $n$ = number of years of record
$m$ = rank or position of any individual discharge in the series

5. Using these values, construct a graph on semilog graph paper of recurrence interval versus discharge (or stage). Connect the data points with a line of best fit.

It is best to view flood hazard statistics in terms of probability rather than recurrence interval (table 10.1). One value is actually the reciprocal of the other. A 50-year flood, for example, has a 1/50 or 2% chance of occurring in any given year; a 25-year flood has a 4% chance. Always keep in mind that flood predictions are only probabilities. Two 50-year floods can occur in successive years, followed by two centuries during which no 50-year floods occur at all. The predictive graph gives no guarantees for any particular year; rather, it offers only averages based on past occurrences.

## TABLE 10.1

### Likelihood of Floods of Different Magnitudes

| Chance (%) of at Least One Flood of at Least this Size in a Certain Number of Years | | | | | Return Period, Years |
|---|---|---|---|---|---|
| One hundred years | Fifty years | Twenty-five years | Ten years | Any one year | |
| | | | | 50 | 2 |
| | | | | 40 | |
| | | | | 30 | |
| | | | | 25 | |
| | | | | 20 | 5 |
| | | 99 | 80 | 15 | |
| | 99.9 | 94 | 65 | 10 | 10 |
| | 90.5 | 71 | 40 | 5 | 20 |
| 86 | 63 | 40 | 18 | 2 | 50 |
| 63 | 39 | 22 | 9.6 | 1 | 100 |
| 39 | 22 | 12 | 5 | 0.5 | 200 |
| 18 | 9.5 | 5 | 2 | 0.2 | 500 |
| 9.5 | 4.8 | 2.5 | 1 | 0.1 | 1000 |
| 5 | 2.3 | 1.2 | 0.5 | .05 | 2000 |
| 2 | 1.0 | 0.5 | 0.2 | .02 | 5000 |
| | 0.5 | .25 | 0.1 | .01 | 10,000 |

*From B. M. Reich,* Water Resources Bulletin, *9:187, 1973. Copyright © 1973 American Water Resources Association. Bethesda, MD. Reprinted by permission.*

Depending on the number of years for which records exist, the flood-frequency curve can be extrapolated, with varying degrees of certainty, to include larger, infrequent events. For example, few streams have records spanning 100 years, so we must often extend plots beyond the data to estimate the magnitude of 100-year events and locate the 100-year floodway. Another factor that leads to uncertainty in flood-frequency curves and recurrence-interval computations is the occurrence of a very large flood (say the 100-year event), within a short period of record (for instance, 20 years); it diverges from all the other data points.

### *Dams and Flood Prevention*

Floods have been a continuing hazard to human habitation since our earliest days on planet Earth; they can never be completely controlled. Man-made structures such as dams can provide a high degree of flood protection. We can also devise plans to avoid the potentially catastrophic effects of a 50-year flood. But is the public willing to bear the cost of protection against the 100-year or 200-year flood? Complicating the picture is the fact that the lake or reservoir that forms behind a dam tends to fill with sediment carried by the inflowing stream. This water-filled, topographically low area behind the dam traps sand and mud that was once carried downstream—sediment that can now be removed only by draining and dredging the lake. The effective life of a dam constructed for flood control, power generation, or recreation often turns out to be much shorter than anticipated. For example, the Tarbela Dam in Pakistan, completed in 1975, is the world's largest earth and rockfill dam, but sediment accumulation will render it useless in less than 50 years. The Anchicaya Dam in Columbia was completed in 1955; 21 months later the lake behind it was already 25% filled with sediment. In Taiwan, mud decreased the capacity of the Shihmen Reservoir by more than 45% between 1963 and 1968. Clearly, for each of these structures the erosion rate of rocks in the drainage basin was not thoroughly understood. Inadequate environmental-geologic studies of dammed regions can prove very costly to the generation that builds the dam, and to succeeding generations as well.

Dams have many effects on downstream locations, including decreasing natural variability in streamflow caused by seasonal and annual fluctuations in precipitation. We can quantify downstream effects in terms of the amount of sediment the stream carries and the changes in the dimensions of the stream channels. These effects, in turn, influence vegetation, wildlife, and human habitation downstream.

Because dams are built for various purposes, they vary widely in the magnitude and duration of their flow releases. Some dams withhold almost all the water from the downstream section. Most such dams were constructed only for recreational and irrigation purposes; they are not needed for power generation, and all the water needed for irrigation is withdrawn from the reservoir. A balance between water inflow to the reservoir and usage and evaporation from it maintains a fairly constant water level. Only occasionally does the level rise so high that water must be released through the dam. At other dams large quantities of water might be diverted during the irrigation season, but all flows (and some sediment) might pass directly through the dam during the rest of the year. Whatever the pattern of controlled releases, they are certain to be distributed differently from the natural flows.

### *Flood-Loss Reduction*

Several steps can be taken to decrease the effects of flooding on human populations. The most important is to educate the public about the frequency and dangers of local flooding. Without an effective educational program, people will not take the steps necessary to ensure safety and minimum loss of life and property.

Many governments have passed regulations concerning floodplain use and occupancy. Foremost among these are zoning laws that limit or regulate the types of construction permitted in specific areas adjacent to stream courses. Such legislation generally requires agreement between real estate developers who own the land and want to maximize their profits from it, and the city or state government, whose main concerns might be different. Zoning laws often are accompanied by building codes, or construction standards that include floodproofing requirements such as placing shields around buildings, or erecting buildings on stilts that raise the bottom floor to several feet above ground level.

Many insurance companies sell flood insurance, and a community can opt to require its purchase by those who insist on owning structures in flood-prone areas. Some communities offer tax incentives for investing in ways to reduce flood loss. However, many Americans dislike programs that force individuals to protect themselves against flooding; instead, they believe the government's responsibility lies in providing relief programs or subsidies, such as low-interest loans, if a catastrophe occurs. Numerous disaster-relief programs now exist at both state and federal levels.

## Problems

1. In 1979 Houston, Texas, had three 100-year floods. What is the probability of this occurrence? What does this do to your confidence about building your house near a river?
2. Based on the changes of the bounding curve (figure 10.1B) in the graph of the largest U.S. flood discharges versus drainage basin area, do you think the position of the most recent curve is likely to change significantly if more recent data (post-1965) are considered? Explain.
3. Obviously building a house close to a river is risky. But how near is "close"? Suppose that the cost of a house decreases as the construction site gets closer to a stream channel. Describe how you could decide on acceptable cost versus probable flood frequency. Flood insurance is prohibitively expensive on the income of a neophyte environmental geologist.
4. The accompanying table (table 10.2) shows mean annual discharges for the Chikaskia River near Blackwell, Oklahoma, over a 38-year period. Rank the discharges and calculate the recurrence interval for each year, then plot the data on the log-log graph paper provided. Eyeball a best-fit line through the 38 data points.
   a. Compare the discharge indicated by your line with that determined by your two nearest neighbors in class, with respect to the 2- and 100-year recurrence intervals. Does your agreement differ for the 2-year versus the 100-year estimates? Would you expect a difference? Why or why not?
   b. Suppose you had plotted the highest 1-month mean discharges for each year rather than the mean discharge for each 12-month period. Do you think the line for the 1-month data would be above or below the mean for each year? Explain. Suppose you plotted the highest 1-day discharges? One-hour discharges? What conclusions do you draw from thinking about these different approaches?
5. Black Bear Creek flows west to east through the center of the Garber Quadrangle (see fold-out map at back of book). After months of searching you find a site on which to build your dream house, and the current owner is willing to sell the property at a price you can afford. The area is the SE 1/4 of section 20, T22N R3W. However, as an environmentally knowledgeable person you are concerned about the possibility of flooding. You contact the State Water Resources Board and obtain the rating curve (the relationship between stream discharge and river stage) as well as discharge recurrence data for the creek at a gaging station alongside the property (figures 10.4, 10.5).
   a. Outline on the fold-out map (figure 10.6) the drainage basin of Black Bear Creek. How many square miles of the drainage basin of Black Bear Creek are present upstream from the property?
   b. How frequently can you expect the creek level to reach your front door, which will be at an elevation of 1050 feet?

TABLE 10.2

Mean Annual Discharges for the Chikaskia River near Blackwell, Oklahoma, for 1937–1974

| Rank | Year | Mean Discharge ($ft^3/s$) | Recurrence Interval (years) |
|---|---|---|---|
| | 1937 | 266 | |
| | 1938 | 419 | |
| | 1939 | 159 | |
| | 1940 | 76.6 | |
| | 1941 | 199 | |
| | 1942 | 690 | |
| | 1943 | 282 | |
| | 1944 | 694 | |
| | 1945 | 884 | |
| | 1946 | 228 | |
| | 1947 | 687 | |
| | 1948 | 728 | |
| | 1949 | 1170 | |
| | 1950 | 375 | |
| | 1951 | 1450 | |
| | 1952 | 254 | |
| | 1953 | 109 | |
| | 1954 | 71.0 | |
| | 1955 | 307 | |
| | 1956 | 168 | |
| | 1957 | 979 | |
| | 1958 | 420 | |
| | 1959 | 468 | |
| | 1960 | 908 | |
| | 1961 | 656 | |
| | 1962 | 535 | |
| | 1963 | 184 | |
| | 1964 | 189 | |
| | 1965 | 966 | |
| | 1966 | 97.6 | |
| | 1967 | 158 | |
| | 1968 | 337 | |
| | 1969 | 661 | |
| | 1970 | 439 | |
| | 1971 | 161 | |
| | 1972 | 151 | |
| | 1973 | 1130 | |
| | 1974 | 962 | |

**Figure 10.4**

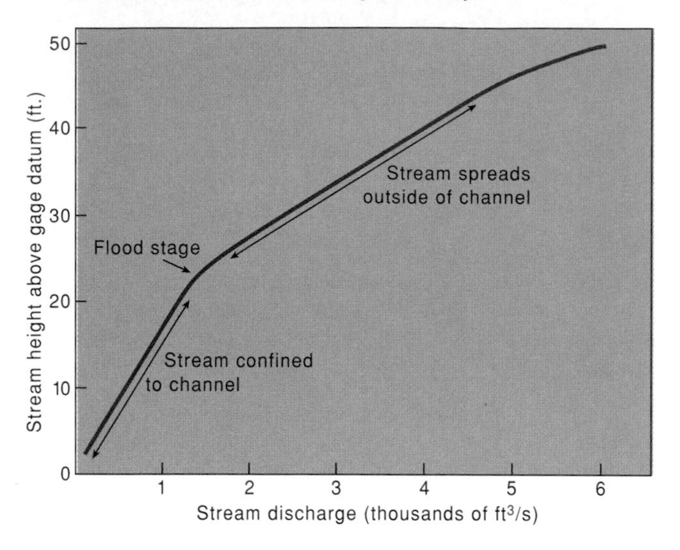


Discharge versus river height for Black Bear Creek in section 20, T22N R3W, Garfield County, Garber Quadrangle, Oklahoma.
*Source: Data from Oklahoma Geological Survey.*

**Figure 10.5**

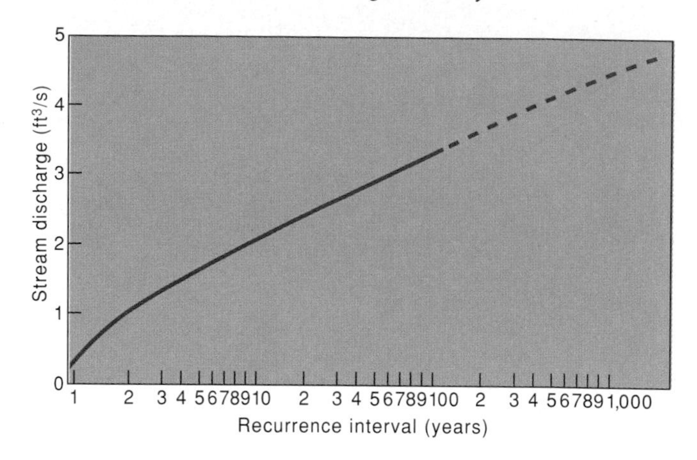


Recurrence interval versus stream discharge for Black Bear Creek in the center of section 20, T22N R3W, Garfield County, Garber Quadrangle, Oklahoma.
*Source: Data from Oklahoma Geological Survey.*

c. Are you willing to build your house there in light of your answer? What frequency of flooding is acceptable to you?
d. Would (or should) your decision about house construction be affected by whether or not a flood has actually attained the 1050 level recently?
e. Should the minor stream tributary on the property be of concern to you? Why or why not?
f. Another potential homesite is the N 1/2 of section 22, T22N R4W. Give two reasons why this might be a better choice from a hydrologic viewpoint.

6. Suppose you are the environmental geologist called to choose the best site for constructing a flood-control dam on a large river. List the factors you think important in selecting the location, briefly explaining the significance of each factor. Consider the lithology and coherence of the rocks around the potential site, the position of the site in the drainage basin, and any other factors you consider potentially significant.

## Further Reading/References

Cudworth, A. G., Jr., 1989, *Flood Hydrology Manual.* Denver, Bureau of Reclamation, U.S. Dept. of the Interior, 243 pp.

Gross, E. M., 1991, The hurricane dilemma in the United States: *Episodes,* v. 14, pp. 36–45.

Gruntfest, E., and Huber, C. J., 1991, Toward a comprehensive national assessment of flash flooding in the United States: *Episodes,* v. 14, pp. 26–35.

Leopold, L. B., 1968, *Hydrology for Urban Land Planning.* U.S. Geological Survey Circular 554, 18 pp.

Wahlstrom, E. E., 1974, *Dams, Dam Foundations, and Reservoir Sites.* New York, Elsevier, 278 pp.

Williams, G. P., and Wolman, M. G., 1984, *Downstream Effects of Dams on Alluvial Rivers.* U.S. Geological Survey Professional Paper 1286, 83 pp.

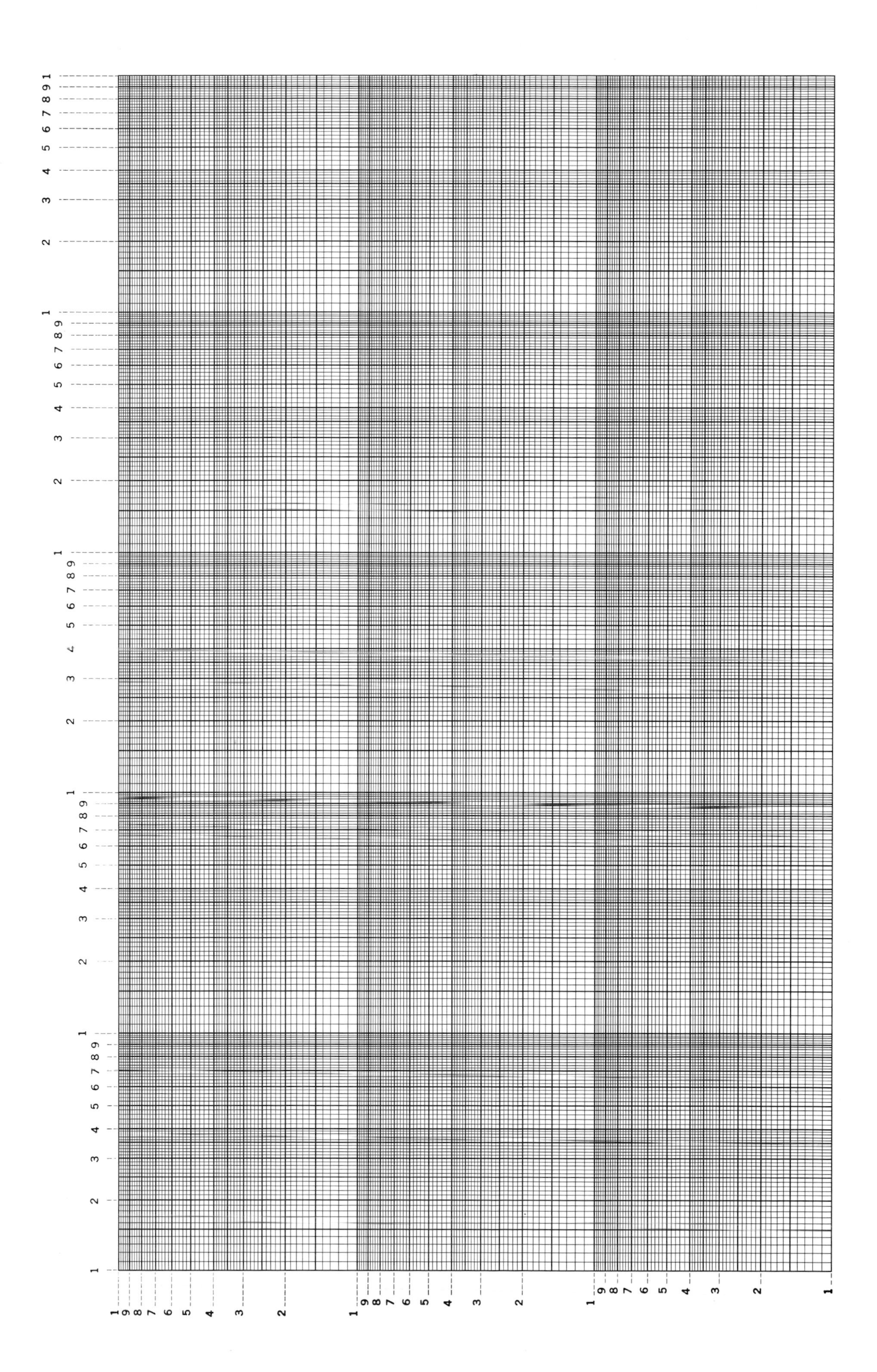

# exercise ELEVEN

## Coastal Erosion

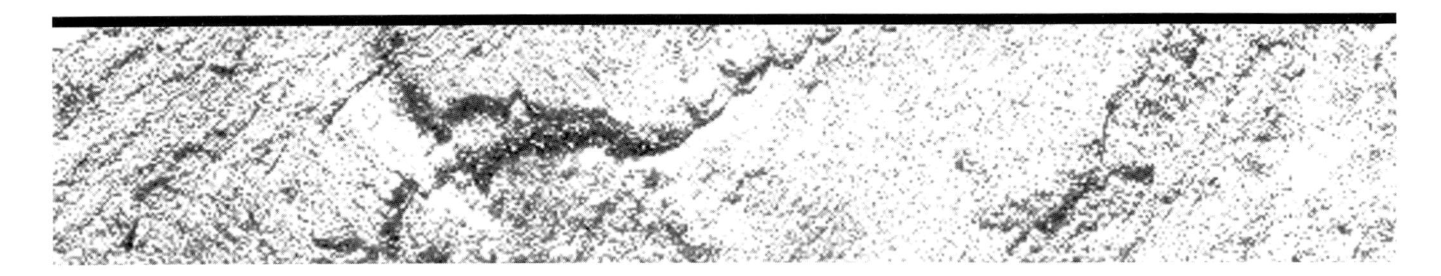

The coastal zone is an area of variable width that includes land and sea areas close enough to the shoreline to be affected by nearshore processes (figure 11.1). In some areas of the United States the coastal zone is tens of miles wide. Processes that affect this zone include waves, currents, hurricanes (figure 11.2), tides, tsunamis, and floods at the mouths of large rivers such as the Mississippi. More than 25% of the American population lives in areas close enough to the shoreline to be affected by at least one of these processes, and the percentage is increasing yearly. This high percentage of nearshore dwellers stems in part from the locations of the seaports where most of the nation's immigrants landed between 1860 and 1920, and in part from the esthetic value of many coastal areas. Coastal communities account for more than half the residential construction at present. By 2010 the coastal population is expected to reach 127 million.

### *Sand Movement*

Human activities continually interfere with natural coastal processes, causing pollution and ecologic changes; conversely, the same natural processes interfere with people's enjoyment of the seashore. Foremost among the natural processes is sand movement and the resultant destruction of beaches and barrier sand bars that have both commercial and esthetic value. Beach sand is lost permanently through transportation into deeper water, beyond the reach of shoreward wave and current action, through transportation by onshore winds to inland dunes, and through pulverization into particles too small to remain on the beach.

The bulk of sediment movement along coastlines results from waves and the longshore current they spawn. Wave energy per $cm^2$ of wave in seawater is calculated by the formula

$$E = \frac{\rho g H^2}{8}$$

$\rho$ = density of the water (1.03 $g/cm^3$)
$g$ = gravitational acceleration (980 $cm/s^2$)
$H$ = wave height (cm)

Wavelength depends on wind strength and duration, the distance over which the wind blows (the fetch), and the configuration of the shore bottom. The water molecules within a wave move in a circular pattern, with a net forward motion toward the shore; when the water depth becomes less than about one-half the wavelength, the circles start to scrape on the sea bottom (figure 11.3). This contact slows the lower part of the water mass, while the upper part rushes forward with an accompanying increase in wave height. Because the energy of the onrushing wave increases as the square of wave height, high waves have very large erosive capabilities. During storms and winter months wave action moves much more sediment offshore than is replaced during nonstorm and summer months.

If the sea-floor surface in the nearshore area remained perfectly smooth as water depth increased and if the wind blew exactly perpendicular to a straight shoreline, then all

**Figure 11.1**

Coastal erosion in the United States. All 30 coastal states are experiencing erosion along their coastlines.

*Source: U.S. Geological Survey.*

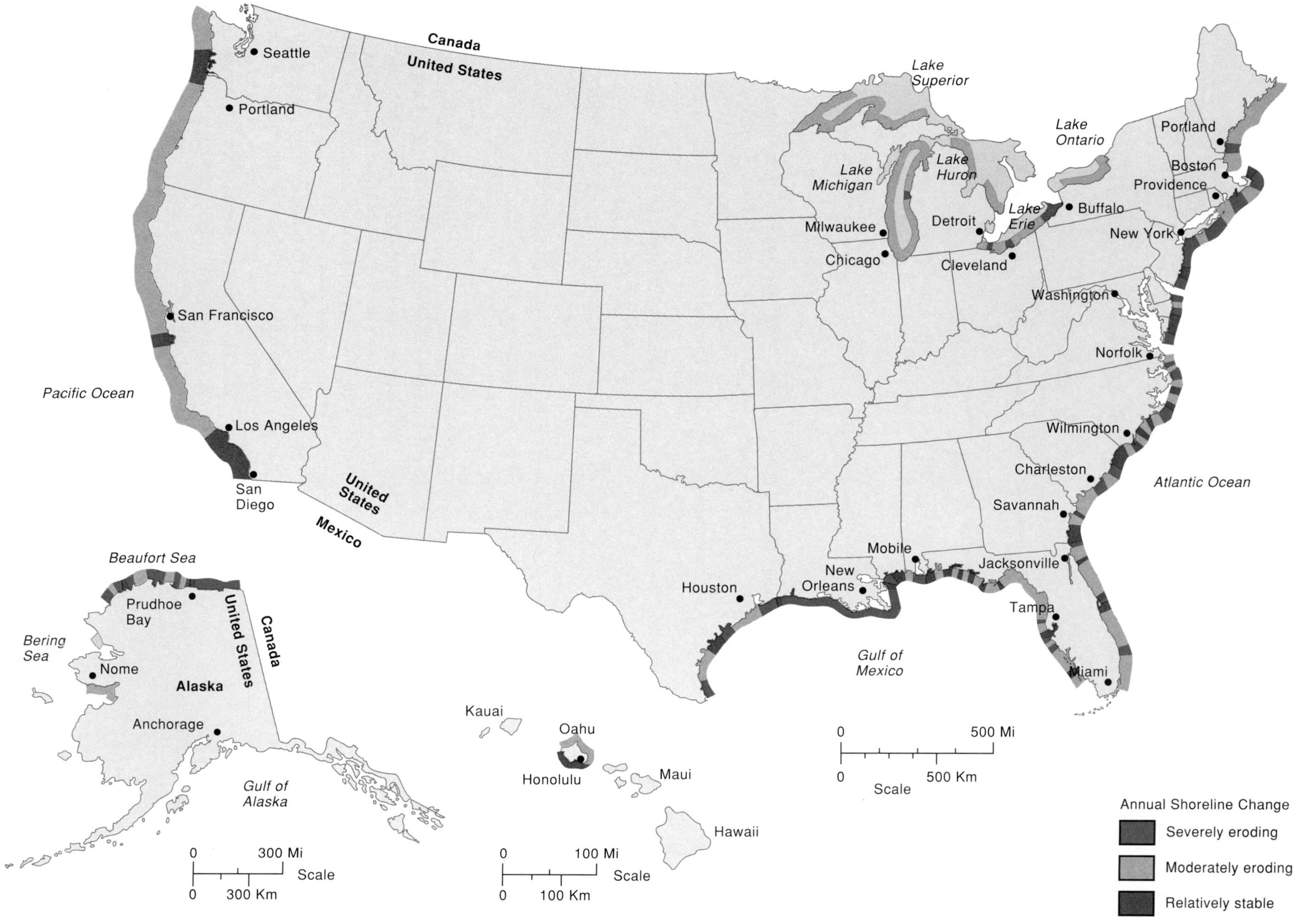

Canada
United States
Seattle
Portland
San Francisco
Los Angeles
San Diego
Pacific Ocean
United States
Mexico
Lake Superior
Lake Michigan
Lake Huron
Lake Ontario
Lake Erie
Milwaukee
Chicago
Detroit
Cleveland
Buffalo
Portland
Boston
Providence
New York
Washington
Norfolk
Wilmington
Charleston
Savannah
Atlantic Ocean
Jacksonville
Tampa
Miami
Mobile
New Orleans
Houston
Gulf of Mexico
0 500 Mi
0 500 Km
Scale
Beaufort Sea
Prudhoe Bay
Bering Sea
Nome
Alaska
United States
Canada
Anchorage
Gulf of Alaska
0 300 Mi
0 300 Km
Scale
Kauai
Oahu
Honolulu
Maui
Hawaii
0 100 Mi
0 100 Km
Scale
Annual Shoreline Change
Severely eroding
Moderately eroding
Relatively stable

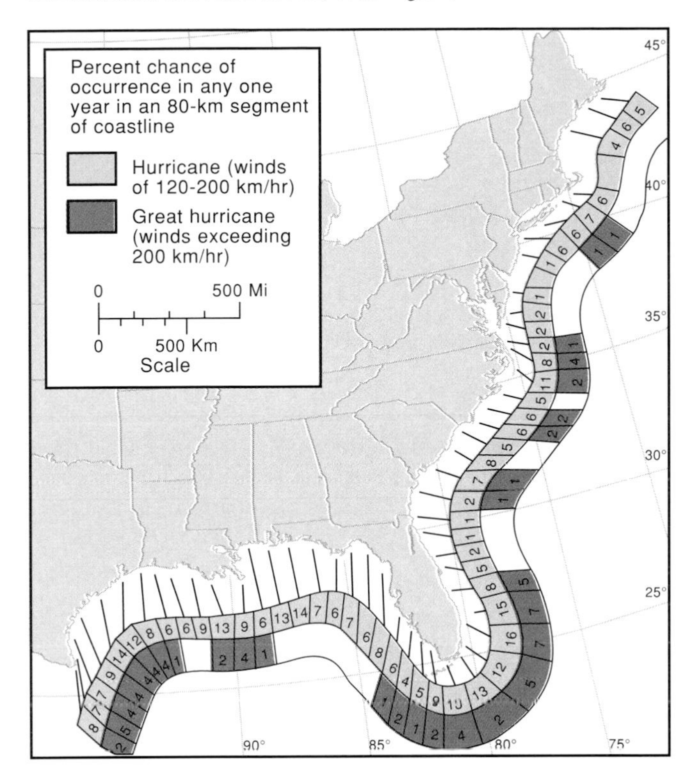


## Figure 11.2

Probability that a hurricane will strike a particular 80km segment of the south Atlantic coast in a given year.

*Source: R. H. Simpson and M. B. Lawrence, 1971, Atlantic hurricane frequencies along the U.S. coastline:* NOAA Technical Memorandum No. NWS-SR-58, *Washington, D.C.*

the energy of the wind-generated waves would be expended normal to the beach. In the real world, however, such a situation never occurs; instead, some wave energy is focused parallel to the beach and generates currents, called *long-shore currents,* that move sediment parallel to the shore (figure 11.4A). The sand-transporting capacity of these currents is given by

$$I = kP$$

$I$ = immersed weight of sand transported per unit time
$k$ = dimensionless constant, averaging 0.77
$P$ = longshore component of wave power (metric units)

The longshore component of wave power, $P$, is related to the power of the waves per unit crest length of breaking waves, $P_1$, by the formula

$$P = P_1 \sin\alpha \cos\alpha$$

where $\alpha$ is the breaker angle (the angle between the breaking wave front and the shoreline; (figure 11.4b). For a sandy shoreline, if we can estimate the budget of incoming wave energy, we can use these two equations to estimate long-shore transport of sediment. Measurements on the east and west coasts of the United States give a range of transport volumes from $10^4$ to $10^6$ $m^3$/yr. Most sand movement occurs

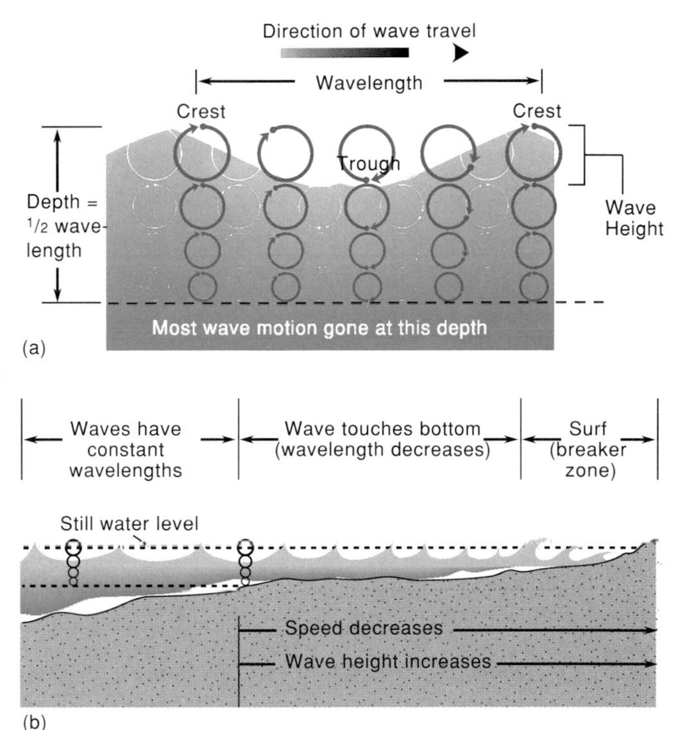


## Figure 11.3

Terminology and behavior of ocean waves in deep water and as the waves approach the shore (beach). (a) Surface winds cause water molecules to move in circular orbits whose diameters decrease with increasing depth. Wave motion is negligible at a depth of 1/2 the wavelength. (b) As the waves move into shallow water—water shallower than 1/2 the wavelength—the orbits of the water molecules become compressed. The moving water scrapes the seafloor, causing erosion and sand movement. Waves increase to a height at which they cannot sustain themselves and collapse to form surf composed of very turbulent water with great erosive power.

in the breaker zone because of the high kinetic energies there. In California, for example, 80% of the longshore drift occurs in water less than 2 m deep. In general, high waves transport sand in the breaker zone; low waves move sand mostly in a zigzag fashion along the beach face.

Along open coasts the currents caused by the twice-daily rise and fall of tides are rarely strong enough to pick up bottom material, although they may transport wave-suspended particles. Tidal range and erosive power are positively correlated. Along open areas of the Atlantic coast the tidal range is only 1–2 m and current strengths are normally less than 1 m/s. In constricted bodies of water such as bays and estuaries, however, the tidal range can be an order of magnitude greater, with a correspondingly greater sand-eroding ability. The world's greatest tidal range occurs in the Bay of Fundy between New Brunswick and Nova Scotia, Canada, where it reaches an astonishing 16.3 m. Current strengths of more than 5 m/s have been measured there.

**Figure 11.4a**

Wave refraction changes the wave direction, bending the wave so it becomes more parallel to shore. The angled approach of waves to shore sets up a longshore current parallel to the shoreline.

*From Charles C. Plummer and David McGeary,* Physical Geology, *5th ed. Copyright © 1991 Wm. C. Brown Communications, Inc., Dubuque, Iowa. All Rights Reserved. Reprinted by permission.*

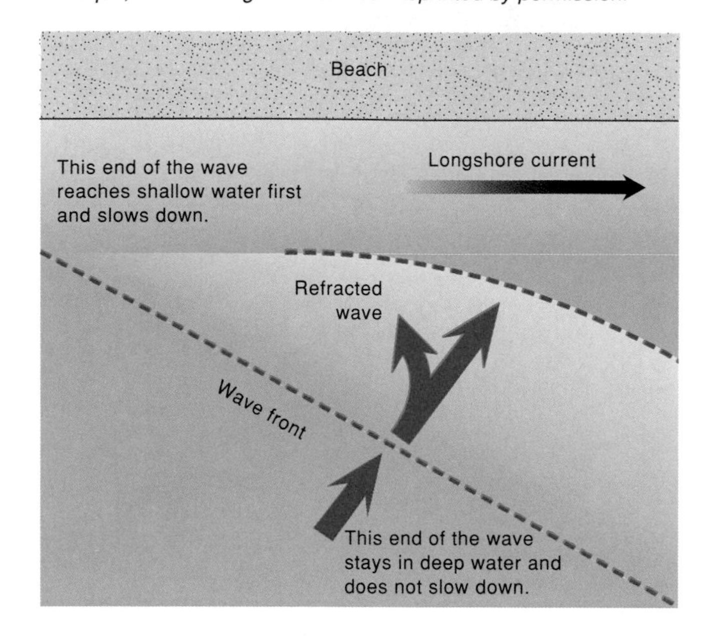


**Figure 11.4b**

The shoreline geometry that controls sand transport.

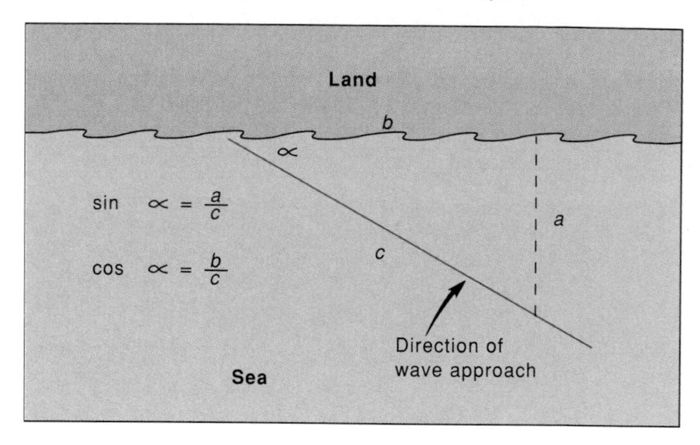


*Tsunamis* or seismic sea waves are generated by earthquakes beneath the ocean floor, by submarine landslides, or by volcanic action. Sea-floor shaking generates water waves with very long wavelengths (sometimes over 500 km), wave heights of a few meters, and periods of 10 to 60 minutes. The velocity of a tsunami wave is given by

$$V = \sqrt{gd}$$

$V$ = wave velocity (m/s)
$g$ = gravitational acceleration (9.8 m/s$^2$)
$d$ = water depth (m)

For the average ocean depth of about 4,000 m, the equation yields mean tsunami velocities of about 200 m/s (450 mph). The largest tsunamis have open-ocean heights of 3–5 m and wavelengths of up to 1,000 km. Wave refraction (bending) in shallow water can transform a 5 m tsunami wave moving at 170 m/s in deep water to a wave 30 m (100 ft) high moving at 15 m/s (33 mph) at a coastline. The erosive effect of such a monstrous wave can be devastating to the coastline and to inhabitants for a considerable distance inland. The tsunami that destroyed Lisbon, Portugal, in 1755 killed 60,000 people; the Krakatoa tsunami in 1883 killed 36,000.

Landward movement of the shoreline is also caused by sea-level rise, now occurring at a rate of about 30 cm/100 yr along the U.S. Atlantic coast. The rise results from the melting of Antarctic glacial ice. Even greater encroachment by the sea occurs in parts of the tectonically sinking Mississippi delta, where the rate of rise can exceed 1.2 m/100 yr (figure 11.5). Neither the melting of glaciers nor the sinking of the delta sediment is caused by human activities; however, the hypothesized increase in the greenhouse effect caused by our dumping of carbon dioxide into the atmosphere might accelerate the rate of sea-level rise by increasing the melting rate of Antarctic ice. Human activity clearly has accelerated shoreline retreat in areas such as Galveston, Texas, and Long Beach, California, where removal of petroleum and water from the subsurface has caused the land to subside and the sea to overrun low-lying land. In some areas river-damming has accelerated inland migration of beaches by trapping sand that would otherwise have been carried to the coast to replenish the beach.

## *Coastal Engineering*

In a continuing effort, engineers have developed two approaches to diminishing or halting the inland movement of a shoreline. One approach includes the construction of seawalls, jetties, groins, and breakwaters. Usually these structures are built of cement and concrete, are anchored to

## Figure 11.5

Effect of 125 years of sea-level rise in the Mississippi delta. Rapid erosion of the protecting barrier islands has exposed Louisiana's valuable wetlands and estuaries to increased storm waves and currents.

*Source: S. J. Williams, et al., 1990, p. 23, U.S. Geological Survey.*

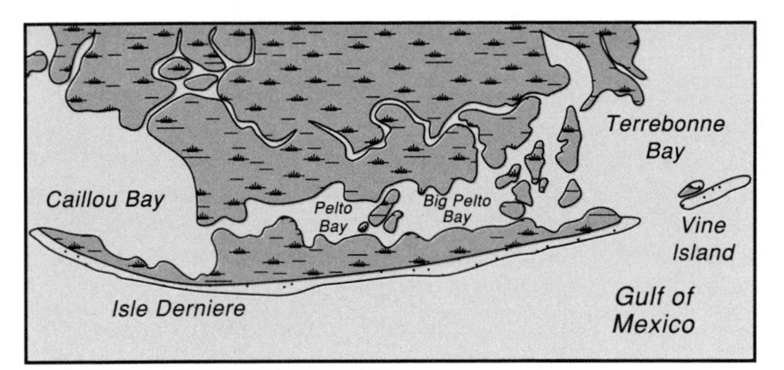


1853

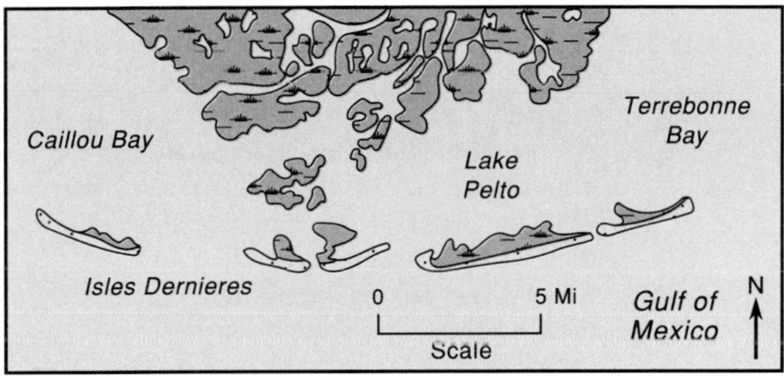


1978

## Figure 11.6

Major types of engineering structures for reducing beach erosion and retreat. In all cases, accumulation of sand in one locality is balanced by sand removal (erosion) in an adjacent locality. Moving water has both an erosive capability and a sediment-transporting capacity that cannot be canceled by artificial structures.

*From William H. Dennen and Bruce R. Moore,* Geology and Engineering. *Copyright © 1986 Wm. C. Brown Communications, Inc., Dubuque, Iowa. All Rights Reserved. Reprinted by permission of the authors.*

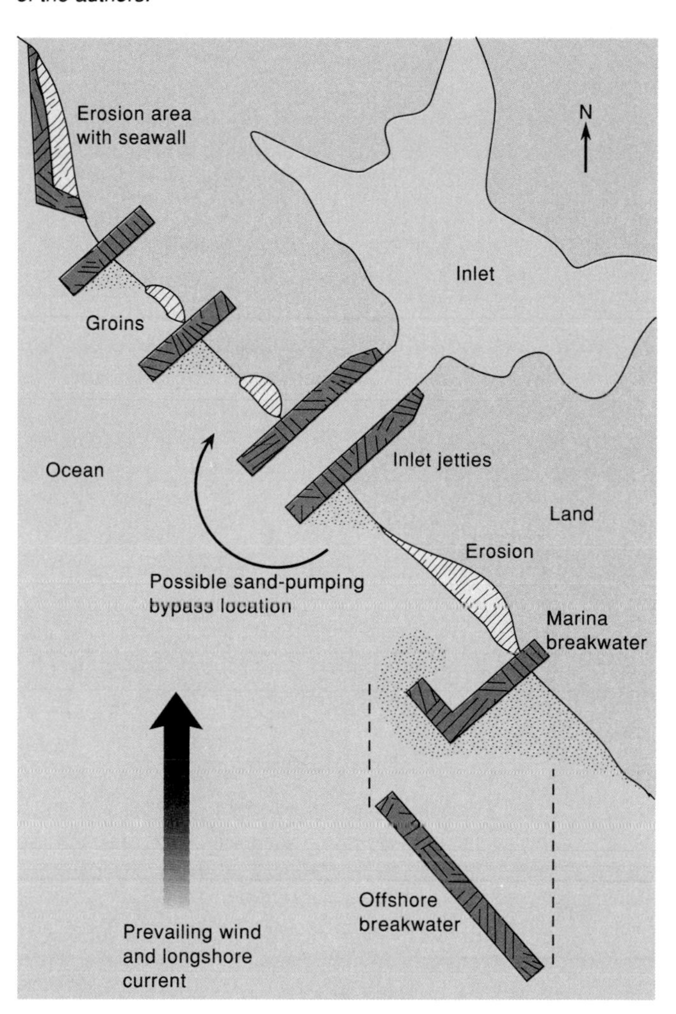


the land and/or shallow sea floor, and are designed to decrease or deflect the wave energy as it approaches the beach. Unfortunately, these structures typically create as many problems as they solve (figure 11.6). Sand moving along the shoreface stops and accumulates on the upcurrent side of the groin or jetty, starving the beach on the downcurrent side. The beach downcurrent erodes as a result. Seawalls built at the inner edge of a beach to protect expensive buildings such as beachfront homes and resort hotels cause erosion of the very beach sand that makes the area desirable; eventually the seawalls are undermined and disintegrate.

The second approach engineers use to maintain beaches is sand replenishment—adding "new" sand to rebuild beaches that have retreated to positions near seawalls or buildings. Sand is usually pumped to the beach from inlets, tidal delta shoals, or the continental shelf; in some cases, it is trucked from inland quarries. Beach replenishment has become more common in recent years because it does not disrupt natural processes, it is a buffer against coastal erosion, and it supplies sand to adjacent beaches. Unfortunately, it is also very costly and seldom lasts very long. The shortest-lived replenishment project of recent years occurred at Ocean City, New Jersey, where in 1982 storms destroyed a 5.2-million-dollar beach in only 2.5

months. As Pilkey (1989) pointed out recently "predictions of beach durability are always wrong, and nobody in the engineering community looks back to evaluate the success or failure of past projects, so no progress has been made in understanding beach replenishment."

### *Coastal Erosion and Public Policy*

During the past 25 years, state governments have become increasingly involved with property owners in such areas of public interest as coastal wetlands and shorelines. Many states have passed laws restricting development of such properties, and often these laws are so restrictive that the owners suffer great financial loss. Should the owners be financially compensated? Some states claim that because property rights are not absolute rights, the state can use its police power to restrict property use to protect public health, safety, and welfare. Under this doctrine, the state is not required to pay compensation to the property owner. Property owners, on the other hand, say that the Fifth Amendment to the United States Constitution forbids the government from confiscating private property, and that declaring property unfit for commercial exploitation is equivalent to confiscation. Under this doctrine, the owners must be compensated for any financial loss the new laws cause.

The government says it cannot afford to pay for the property, but that its police power to regulate use of the property is needed to protect the environment and public safety. Property owners argue that they cannot bear the entire financial burden of taking the property out of circulation. The courts are currently considering this issue.

## Problems

1. Draw a sketch of a coastline, with headlands, and show the change in shape of the wave front as it passes from deep water into shallow water near the headlands. Explain the reason for the change.
2. From a physics viewpoint, why do you think wave height is so important in determining wave energy? How does this relate to surfing along the California coastline?
3. The amount of sand transported by longshore currents is a function of the product of the sine and cosine of the breaker angle (figure 11.4B). Construct a graph showing breaker angle versus the numerical value of this product. At which angle(s) is the longshore component of wave power the greatest and the least? Explain.
4. When examining a topographic and bathymetric map of a sandy coastline, how many ways can you think of to determine the direction of longshore current movement? In other words, what features on the ground or on the shallow sea floor might indicate current patterns?
5. Figure 11.7 is a map showing the coastline about 30 miles north of Seattle, Washington. Read the legend carefully so you understand what the various symbols and colors indicate.
    a. Explain why the ratio between the linear-retreat rate and the volume of sediment eroded is so variable along the coastlines on the map. Should a site where the land is retreating more rapidly always yield a greater volume of sediment?
    b. Why is there an extensive beach at Admiralty Bay?
    c. Why is there a zone of intense wave erosion around the Lake Hancock Target Range but not at Admiralty Bay?
    d. What has caused the extension of land at Point Wilson? Do you think this extension will eventually seal off Admiralty Inlet? Explain.
    e. Explain why wave erosion tends to be heavy for many miles on either side of McCurdy Point. How does this tendency relate to the existence of Beckett Point?
    f. Why is erosion much more intense along the coast just south of Irondale than immediately across the bay at Jorgenson Hill?
    g. A large amount of sediment transport seems to be occurring in this map area. What can you infer about the character of the coastal sediments that form the source materials?
    h. Based on the varying directions of sediment transport shown, the current patterns appear complex. What factors might explain this complexity?
    i. Based on the topographic variations and the patterns of erosion and deposition, select the safest places to build a house. Now choose the least safe places.
6. You are considering buying some beachfront property but are concerned about erosion. What coastline features (rocks, sediments, and ocean) would be important to weigh in evaluating the durability of your property?
7. You own a stretch of seafront property, and some of your neighbors upcurrent are considering constructing a groin to widen the beach in front of their houses. What would your reaction be? What could you do if you could not reach an amicable agreement with them?
8. Beaches and offshore barrier sand bars are desirable places for summer homes and luxury vacation resorts. The federal government provides low-cost loans and insurance against the effects of hurricanes and other disasters in these esthetically pleasing areas. What is your view of this federal policy?
9. What do you think local and state governments should do about protecting seafront areas for general recreational purposes?

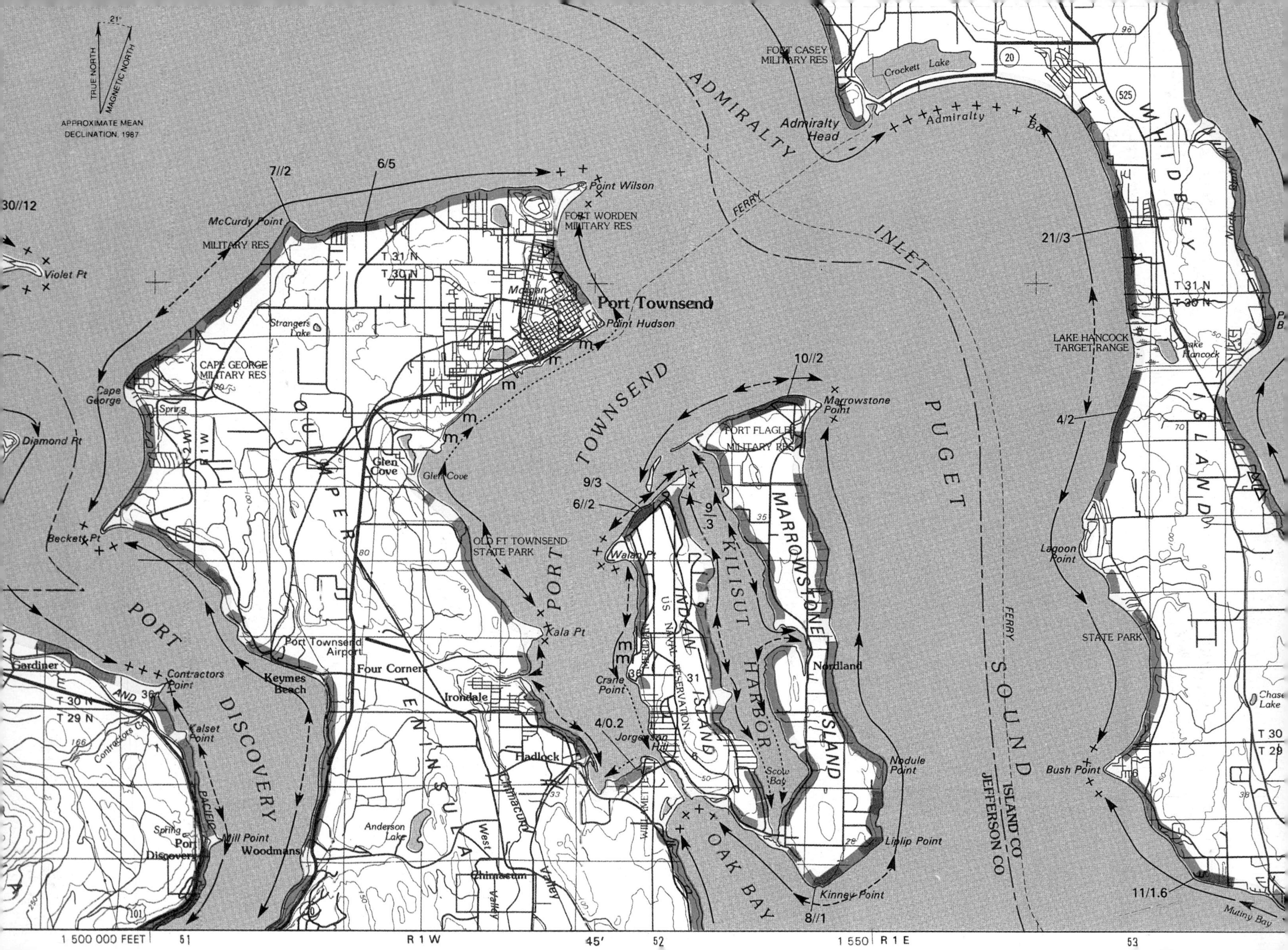

TRUE NORTH
MAGNETIC NORTH
21°
APPROXIMATE MEAN DECLINATION, 1987
ADMIRALTY INLET
PUGET SOUND
FERRY
ISLAND CO
JEFFERSON CO
WHIDBEY ISLAND
FORT CASEY MILITARY RES
Crockett Lake
Admiralty Head
Admiralty Bay
21//3
LAKE HANCOCK TARGET RANGE
Lake Hancock
4/2
Lagoon Point
STATE PARK
Bush Point
Chase Lake
11/1.6
Mutiny Bay
7//2
6/5
30//12
McCurdy Point
MILITARY RES
Point Wilson
FORT WORDEN MILITARY RES
Port Townsend
Point Hudson
Strangers Lake
CAPE GEORGE MILITARY RES
Cape George
Violet Pt
Diamond Pt
Beckett Pt
Glen Cove
QUIMPER PENINSULA
PORT TOWNSEND
OLD FT TOWNSEND STATE PARK
Kala Pt
Port Townsend Airport
Four Corners
Irondale
Hadlock
Anderson Lake
Chimacum Valley
West Valley
PORT DISCOVERY
Keymes Beach
Gardiner
Contractors Point
Kalset Point
Mill Point
Woodmans
Port Discovery
10//2
Marrowstone Point
FORT FLAGLER MILITARY RES
MARROWSTONE ISLAND
Nordland
Nodule Point
Liplip Point
Kinney Point
8//1
KILISUT HARBOR
Scow Bay
INDIAN ISLAND
US NAVAL RESERVATION
9/3
6//2
Walan Pt
Crane Point
4/0.2
Jorgenson Hill
OAK BAY
1 500 000 FEET
51
R 1 W
45′
52
1 550
R 1 E
53

## Further Reading/References


Dolan, R., Anders, F., and Kimball, S., 1988, *Coastal Erosion and Accretion—National Atlas of the United States*. Reston, Virginia, U.S. Geological Survey, 1 sheet.

Dolan, R. and Lins, H., 1985, *The Outer Banks of North Carolina*. U.S. Geological Survey Professional Paper 1177-B, 103 pp.

Fletcher, C. H., III, 1992, Sea-level trends and physical consequences: Applications to the U.S. shore. *Earth-Science Reviews,* v. 33, pp. 73–109.

Living with the Shore series. Durham, North Carolina, Duke University Press. (A continuing series of books for lay people, each book dealing with a different location along the U.S. coastline.)

National Research Council, 1990, *Managing Coastal Erosion*. Washington, D.C., National Academy Press, 182 pp.

Pilkey, O. H., 1989, The engineering of sand: *Journal of Geological Education,* v. 37, pp. 308–311.

Pilkey, O. H., 1991, Coastal erosion: *Episodes,* v. 14, pp. 46–51.

Thieler, E. R., and Bush, D. M., 1991, Hurricanes Gilbert and Hugo send powerful messages for coastal development: *Journal of Geological Education,* v. 39, pp. 291–298.

Walker, H. J., 1988, *Artificial Structures and Shorelines*. Dordrecht, the Netherlands, Kluwer, 708 pp.

Williams, S. J., Dodd, K., and Gohn, K. K., 1990, *Coasts in Crisis*. U.S. Geological Survey Circular 1075, 32 pp.


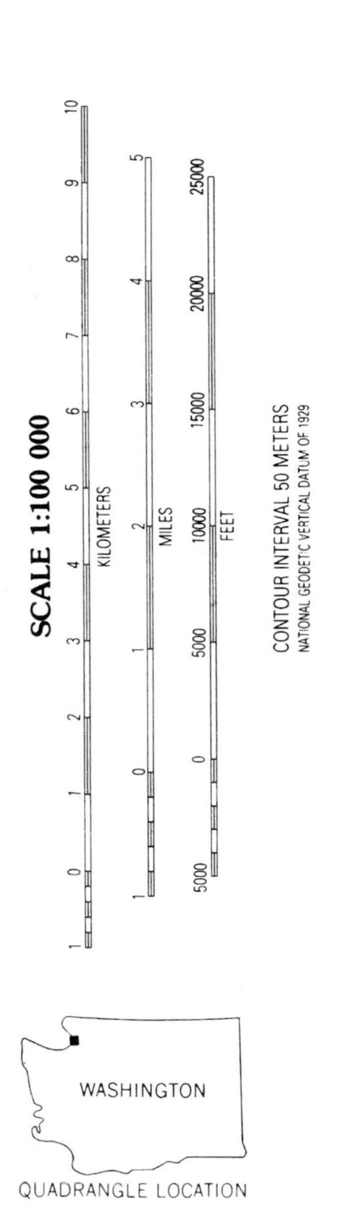


### EXPLANATION

ZONE OF SUBSTANTIAL WAVE EROSION AND EROSION-INDUCED LANDSLIDES

ZONE OF RELATIVELY SLOW WAVE EROSION—Relatively frequent but small landslides common

APPROXIMATELY NEUTRAL COASTAL SEGMENT—Little or no net erosion or deposition

DEPOSITIONAL BEACH

△ △ △ ZONE OF SUBSTANTIAL SEDIMENT LOSS FROM BLUFFS—Caused by large sporadic landslides (little or no direct wave erosion)

DIRECTION OF LONG-TERM NET-SEDIMENT TRANSPORT

Unimpeded

Sediment volume and (or) rate of movement impeded by natural or manmade conditions

NULL (NODAL) ZONE—Zone where sediment being transported along shore diverges into two adjacent littoral cells on a net long-term basis

CONVERGENT ZONE—Zone where sediment from two drift cells meets

EROSION RATES—First or upper number shows rate of bluff retreat, in centimeters per year. Second or lower number shows volume of bluff material lost, in cubic meters per meter of shoreline per year

10/5 Average erosion rate based on an accurately known retreat distance and at least 20 years of record

10//5 Minimum erosion rate; true rate is likely to be greater. Rate is averaged over at least 20 years of record

$\frac{10}{5}$ Average erosion rate based on less than 20 years of record. Rate shown may be representative of a long-term average but has an equal chance of being less or greater than the long-term average

ROCKY COAST—No beaches or appreciable longshore sediment transport unless specifically indicated with map symbols

p POCKET BEACH ON ROCKY COAST

m MODIFIED SHORELINE—Consisting of dredge spoils, artificial fill, jetties, docks, seawalls, or dikes. No appreciable longshore transport unless otherwise indicated on map

DELTAIC AND ESTUARINE TIDAL FLATS—Composed of mud and muddy sand. The seaward margin shown on the map is the minus 1-meter depth contour (generalized), which is also the edge of the exposed tidal flat during very low tides. Net transport is shown only where beaches border the muddy flats along the landward edge; more commonly, fringing marshes rather than beaches are present

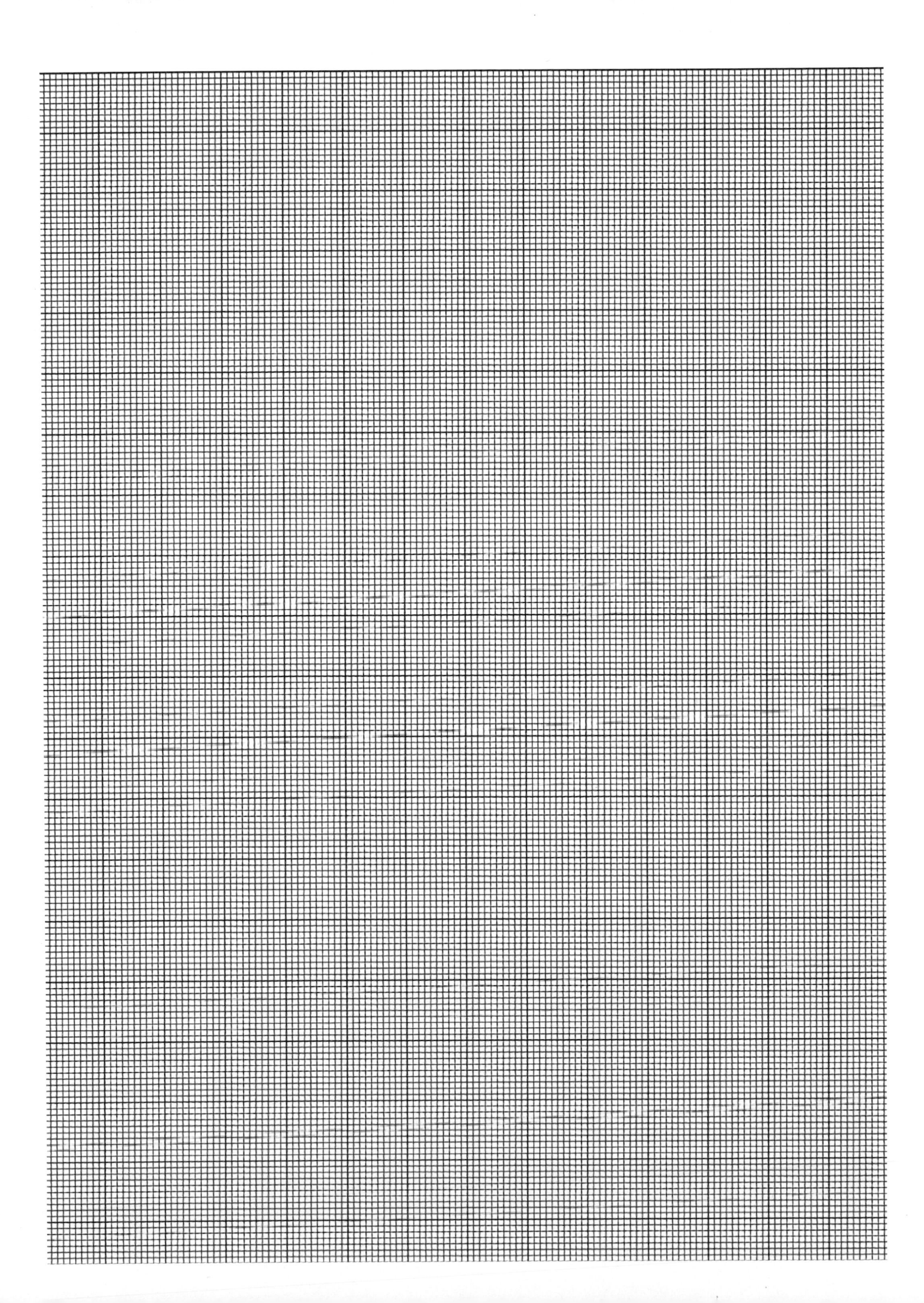

exercise

# TWELVE

## Slope Stability and Mass Movements

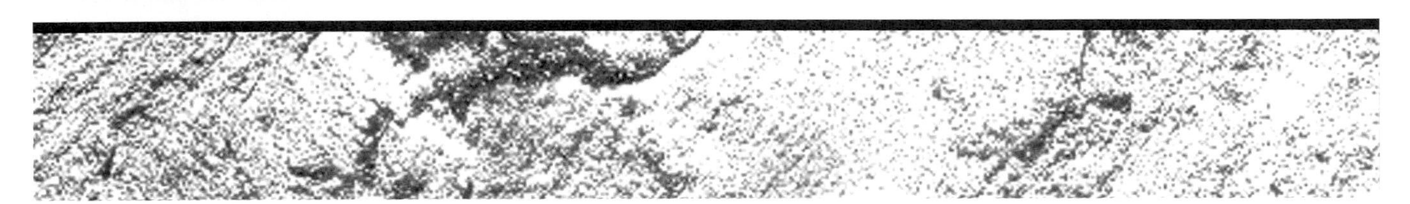

Damage resulting from mass movements of soil and rock is very costly, totalling more than one billion dollars annually in 1978; no doubt, the cost has increased significantly since then. Fortunately, however, mass movements are one of the more controllable natural disasters, given adequate technical knowledge and public awareness.

Hill slopes are formed of natural materials that range in their resistance to downslope movement from the massive bare rock of vertical cliff faces to plastic soils that can flow on even the gentlest slopes. Mass movements occur throughout the entire range of rock and soil properties and slopes. Mass movements are differentiated from streamflow by the ratio of rock and sediment to water; in a mass movement this ratio is very high, ranging from near infinity in an avalanche or rockfall (figure 12.1) to about unity in an earthflow. In streamflow the ratio is less than 0.1, that is, the mass of suspended sediment carried in a stream is less than 10% of the mass of the stream water. Typically it is less than 1%.

### *Causes of Mass Movements*

Hard, intact rocks do not exist in nature; rock strength is controlled by the extent and spacing of discontinuities within the rock mass. Examples of naturally occurring discontinuities in rocks include (1) depositional features such as bedding and fissility, (2) erosional features such as unconformities and scours, (3) metamorphic features such as slaty cleavage and schistosity, (4) tectonic features such as joints and faults, and (5) mineralogic features such as the presence of clay minerals. All these features create mechanical or chemical loss of cohesion between rock surfaces. Because such discontinuities exist everywhere, unstable slopes and mass movements can occur in nearly all terrains and climates. All that is needed for a mass movement to occur is a triggering mechanism strong enough to overcome the rock-rock or rock-soil friction that maintains the existing slope. Naturally occurring triggers include earthquakes, volcanic eruptions, rainfall, snowmelt, and undercutting by stream erosion or sea waves. The resulting movements of soil and/or rock are named according to the amount of liquid water they contain and their cohesiveness. Cohesive movements are generally called earthflows, slumps, or creeps; noncohesive movements may be called avalanches, slides, or rockfalls.

Resistance to ground failure results from two factors. The first is internal friction caused by grain-to-grain contacts between irregularly shaped, coarse granular particles. Any granular material has the ability to stand at some angle or maintain some slope because of grain-to-grain friction. The second factor is cohesion caused by materials that hold particles together in a solid, impermeable mass, such as occurs in clay. Common binding forces are electrostatic attraction or chemical bonds. Shales exemplify this type of cohesion.

## Figure 12.1

Examples of landslides by type of movement.

*Source: W. W. Hays, (editor), 1981, Facing Geologic and Hydrologic Hazards: Earth Science Considerations. U.S. Geological Survey* Professional Paper 1240-B, p. 61.

**Fall (rock fall)**

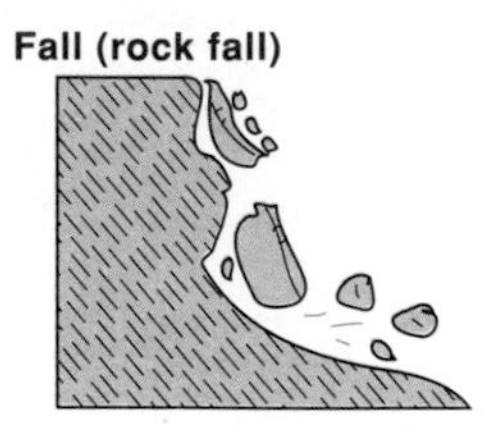

Fall – Mass travels most of the distance in free fall.

**Topple (debris topple)**

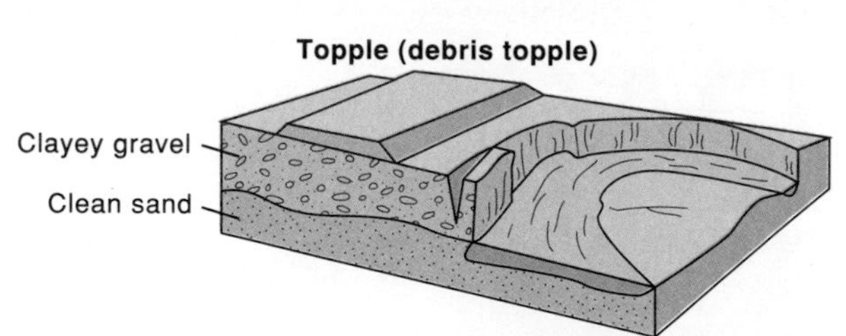


Topple – An overturning movement that, if unchecked, will result in a fall or slide.

Slide – Movement of material by shear displacement along one or more surfaces or within a relatively narrow zone.

**Slide (rock slump)**

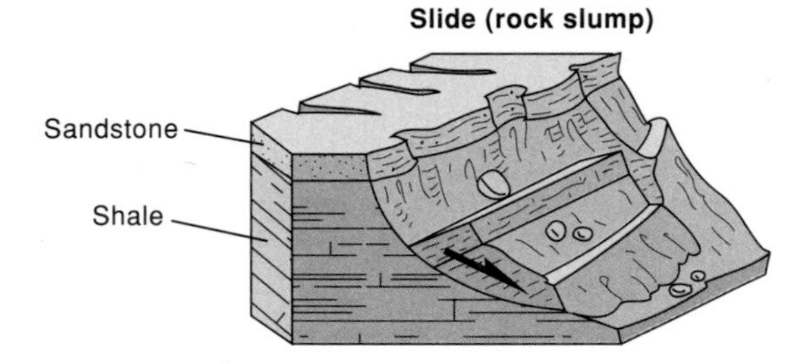


Rotational Slide – Movement involves turning about a point (surface of rupture is concave upward).

**Slide (rock slide)**

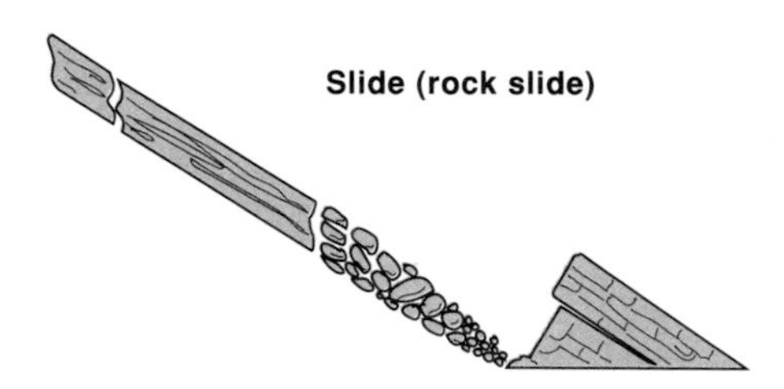

Translational Slide – Movement is predominantly along planar or gently undulatory surfaces, frequently controlled by surfaces of weakness, such as faults, joints, or bedding planes.

**Lateral Spread (earth spread)**

Firm clay

Soft clay with water-bearing silt and sand layers

Firm clayey gravel

Lateral Spread – Lateral extension movement of a fractured mass.

**Flow (earth flow)**

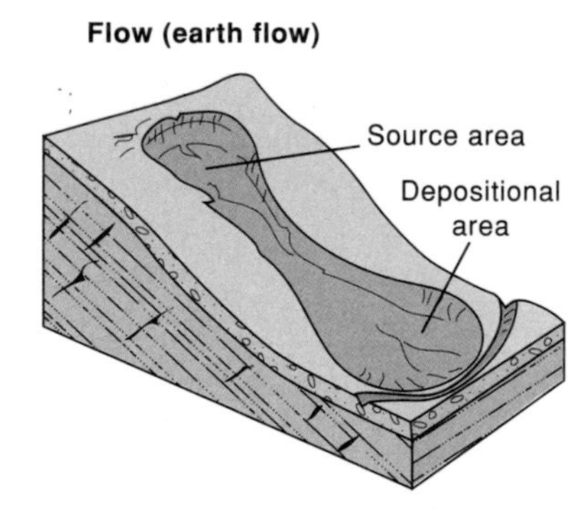


Flow – Movement of mass such that the form taken by moving material resembles that of viscous fluids.

**Complex (slump-earth flow)**

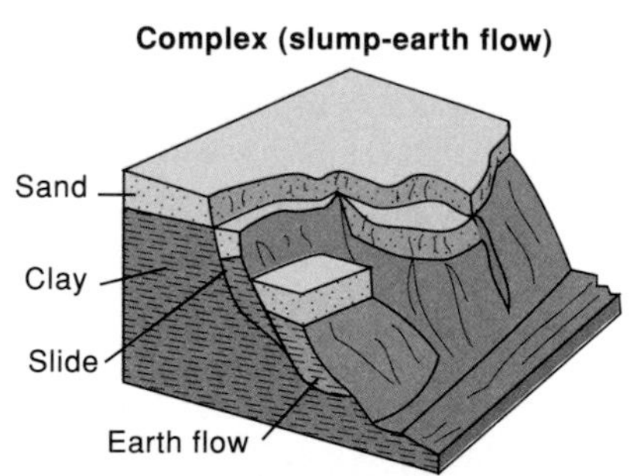


Complex – Landslide incorporating two or more types of movement.

TABLE 12.1

Relative Values for the Major Clay Minerals of Properties Important for the Onset of Slope Failure

| Mineral | Surface Area ($m^2g$) | Liquid Limit | Plastic Limit | Shrinkage Limit | Volume Change | Cation Exchange Capacity (meq/100 g) |
|---|---|---|---|---|---|---|
| Montmorillonite | 800 | High | High | Low | High | 80–120 |
| Illite | 80 | Medium | Medium | Medium | Medium | 10–40 |
| Kaolinite | 15 | Low | Low | High | Low | 3–20 |

The great variability in CEC values results from the effect of variations in chemical composition.

Figure 12.2

Change in volume of $H_2O$ with temperature (not to scale).

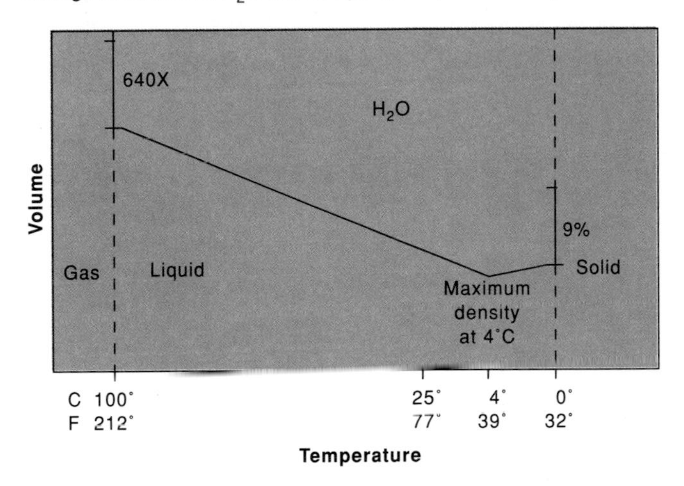


Studies of potential slope failure in coarse granular rocks usually focus on joint spacing. Along joint surfaces the two rock halves make contact only at surface irregularities. These contacts are not like those of an interlocking picture puzzle, so the only force maintaining rock stability and cohesion is grain-against-grain friction. This frictional resistance to slab movement can be decreased by shaking (earthquakes), by water percolation through discontinuities, or by the 9% expansion of water as it freezes (figure 12.2). Percolating water decreases grain-to-grain friction not only by increasing pore pressure but also by causing chemical alteration (weathering) of these surfaces, softening them through the formation of clay minerals. Clay minerals decrease cohesion because they are soft and easily gouged, are sheetlike in character, and absorb water. As the proportion of clay minerals in an aggregate of grains increases, the type of slope failure changes from noncohesive to cohesive.

Most soil failures (earthflows, slumps) are cohesive because most soils contain high percentages of clay. In general, the greater the quantity of clay minerals in a soil, the greater its potential for shrinkage and swelling, the higher its plasticity, and the greater the likelihood that slope failure will be cohesive. Consequently, studies of soil strength and resistance to downslope movement usually focus on clay content. Terms used to describe clay mineral properties in relation to ground failure include *liquid limit, plastic limit,* and *shrinkage limit.* The liquid limit is the smallest amount of clay needed to cause a water-clay mixture to behave as a plastic solid rather than as a fluid. As the clay/water ratio continues to increase, the resistance of the soil to shearing stress increases. Finally the soil reaches its plastic limit, the point at which it exhibits no plastic deformation and fails by brittle fracture. Further decrease in water content causes the soil volume to decrease until it reaches a stable point termed the shrinkage limit. The three limits thus defined are called Atterberg limits, after their originator (table 12.1).

Montmorillonite has a high liquid limit, meaning that plastic behavior begins at low clay contents. It also has a high plasticity limit, meaning that the clay-water mixture maintains plasticity at very high clay/water ratios, a result of large amount of bound water in montmorillonite compared to the other clay types, illite and kaolinite.

Ground failure can also result from processes unrelated to discontinuities in solid rock or to the amount of montmorillonite in clay-rich materials. Two of these processes are *liquefaction* and *spreading failures.* Liquefaction occurs when granular materials change in behavior from solid to liquid in response to increased pore pressure. Groundshaking reorients the unconsolidated sediment grains into a more compact arrangement. If the groundwater table is near the surface during this reorientation, the grain-to-grain contacts are reduced, and the load is temporarily transferred to the pore water. These changes increase pore pressure and decrease strength, and the deposit then behaves as a liquid.

In spreading failures flowage of underlying material causes overlying, firmer rock or soil to break into units and spread apart. The most spectacular failures occur when a fine-grained sediment changes from a fairly hard, strong, brittle solid to a liquid of negligible strength (figure 12.3). Materials exhibiting this unusual property are termed quick clays, although some of these sensitive sediments contain little or no clay. If the sediment is a clay, its clay flakes are arranged in a fluid-filled, house-of-cards structure; when the aggregate is disturbed the structure collapses, and the material behaves like a fluid. If the sensitive sediment is not clay, it is composed of clay-size quartz particles so small that van der Waals bonding forces are large compared to the weight of the quartz grains. Such bonds decrease in effectiveness as the seventh power of the distance between particles, however, so when the aggregate is disturbed and the particles separate slightly, they lose cohesion and commence liquidlike flow.

## *Recognizing and Preventing Slope Failures*

Various studies have shown that the most damaging slope failures are closely related to human activities; regulating land use *before* these activities take place can substantially reduce loss. The old adage "An ounce of prevention is worth a pound of cure" applies well to slope failures. Because scientists and engineers understand the factors that make slopes unstable (figure 12.4), prevention is an attainable goal in many regions. Dangerous areas can be (1) avoided if possible (figure 12.5); (2) stabilized by structures that include retaining walls, drainage ditches, subsurface pipes, and wire mesh coverings; or (3) stabilized by grading slopes or planting trees, whose roots improve soil cohesion.

Approximately 20 years ago the U.S. Geological Survey began engineering-geology mapping projects in the San Francisco Bay area. This region was chosen because it is prone to both earthquakes and landslides and is heavily populated. The results (for example, figure 12.6) are helpful to both engineering geologists concerned with construction and land-planners concerned with a wider range of possible land uses. Since 1970 many state governments have commissioned maps for various environmental purposes such as landslide control, groundwater safety, and soil erosion and pollution. The number and variety of such maps is increasing rapidly as public concern about environmental problems increases.

**Figure 12.3**

Comparison of Leda Clay undisturbed (left), supporting a load of 11 kg, and remolded (right). Water content is identical for both samples.

## Problems

1. Hollidaysburg is a town of 6,000 people in southwestern Pennsylvania near the confluence of two rivers and surrounded by mountains with up to 450 feet of local relief. Because of this location, considerable potential for mass movements of surficial materials exists within five miles of the town. The town government has established a committee to determine

## Figure 12.4

Cross-section showing homeowners with no hope. Their house is built on a combination of unstable fill and shale whose layering is parallel to the ground slope. In addition, the shale and sandstone behind the house are unstable, and the sandstone feeds water into the planes of fissility in the overlying shale; sinkholes are developing in the limestone below the house and will soon undermine the house; and the toe of the fill supporting the house is being undermined by water leaking from the sandstone.

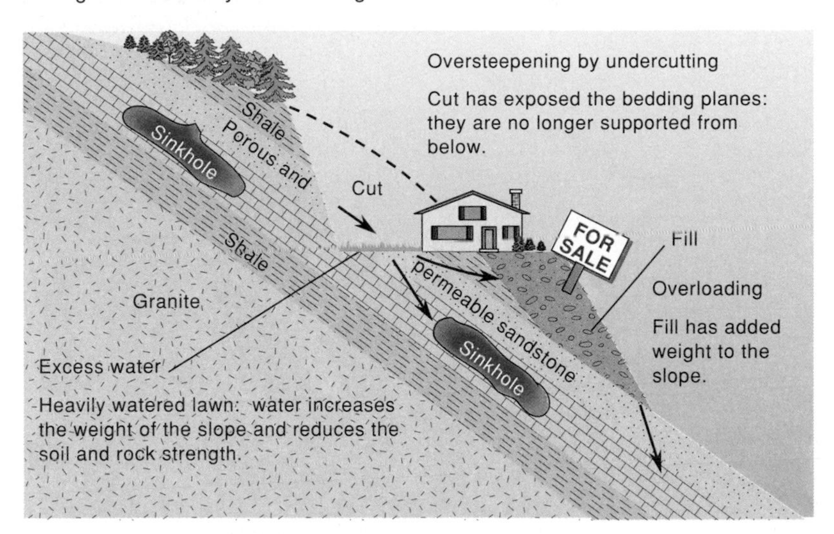


the extent of danger from natural disasters. As a member of this committee and an educated person with some training in environmental geology, you have obtained a geologic and topographic map of the area (figure 12.7). Based on this map, you are asked to answer the following questions:

a. What is the range in geologic age of the rocks exposed in the map area?
b. What type of geologic contact is present between Qal and Dha?
c. What type of sedimentary rock—sandstone, shale, or limestone—is most abundant in the area? Least abundant? What do your answers suggest about the frequency of mass movements in the map area?
d. Identify the formation that should be the most resistant to mass movement. What are the characteristics that caused you to choose it?
e. In what way will the relative proportions of the three rock types affect the useful life of the reservoir located 1.5 miles SSE of the center of Hollidaysburg?
f. What is the ground-slope angle of the hill immediately south of the reservoir? What is the dominant type of rock exposed on the hill?
g. The southwestern part of the city is built on Qal. What environmental problems do you think this might lead to?
h. At the northwest edge of the town of Newry is a steep hill underlain by the formation labeled Db. Do you think the town faces any danger from this hill? Explain.
i. Most of Hollidaysburg is built on Sc and Smk. Might this be a possible problem for the inhabitants located closest to Beaverdam Branch? Explain.
j. Is there any danger to the railroad track from the hill just east of Kladder Station? Explain.
k. Can you suggest one or two reasons why the stream valley is widest about one mile ESE of Newry?
l. What is the general relationship between the dip of the sedimentary rocks, the types of rock, and the propensity for mass wasting?

2. Describe the various ways in which water is involved in slope failures.
3. Discuss the relative effects of rock characteristics and climate on the occurrence of slope failures.
4. List the factors that affect whether an earth movement will be slow, such as an earthflow, or fast, such as an avalanche.

**Figure 12.5**

Aerial photograph showing the result of the most destructive landslide in the history of Hong Kong, June 18, 1972. Almost 26 inches of rain fell during the preceding two days, saturating and undermining the steep slope, which is composed of soil and underlying unlithified coarse sediment. The slope failure was 220 feet wide and destroyed a 4-story building and a 13-story apartment building; 67 people were killed. Hong Kong's population density and topographic relief make it impossible to completely avoid construction in unsafe areas.

5. How might you distinguish between mass-movement deposits and stream or glacial deposits?
6. Volcanoes are commonly sites of massive earth movements. List reasons why the presence of a volcano or group of volcanoes increases the risk of such occurrences.
7. The state highway department is considering constructing a new highway along the base of a hill composed of limestone. What kinds of environmental problems might the construction generate? How should they be investigated and remedied?
8. What should be the various roles of the individual homeowners, the state government, and the federal government in protecting lives and property against destructive mass movements?

## Further Reading/References

Anderson, M. G., and Richards, K. S. (eds.), 1987, *Slope Stability.* New York, John Wiley & Sons, 648 pp.

Brabb, E. E., 1991, The world landslide problem: *Episodes,* v. 14, p. 52-61.

Brabb, E. E., and Harrod, B. L., 1989, *Landslides—Extent and Economic Significance.* Rotterdam, the Netherlands, A. A. Balkema, 385 pp.

Costa, J. E., and Wieczorek, G. F. (eds.), 1987, *Debris Flows/Avalanches: Process, Recognition, and Mitigation.* Reviews in Engineering Geology, v. VII, Boulder, Colorado, Geological Society of America, 239 pp.

## Figure 12.6

(a) Slope-failure map of the Congress Springs area, Santa Clara County, California. (b) A derivative map showing potential ground movement and recommended land-use policies. The original maps are on a topographic map base at a scale of 1 inch to 250 feet (1:3,000).

*Source: U.S. Geological Survey, 1982, pp. 24, 32.*

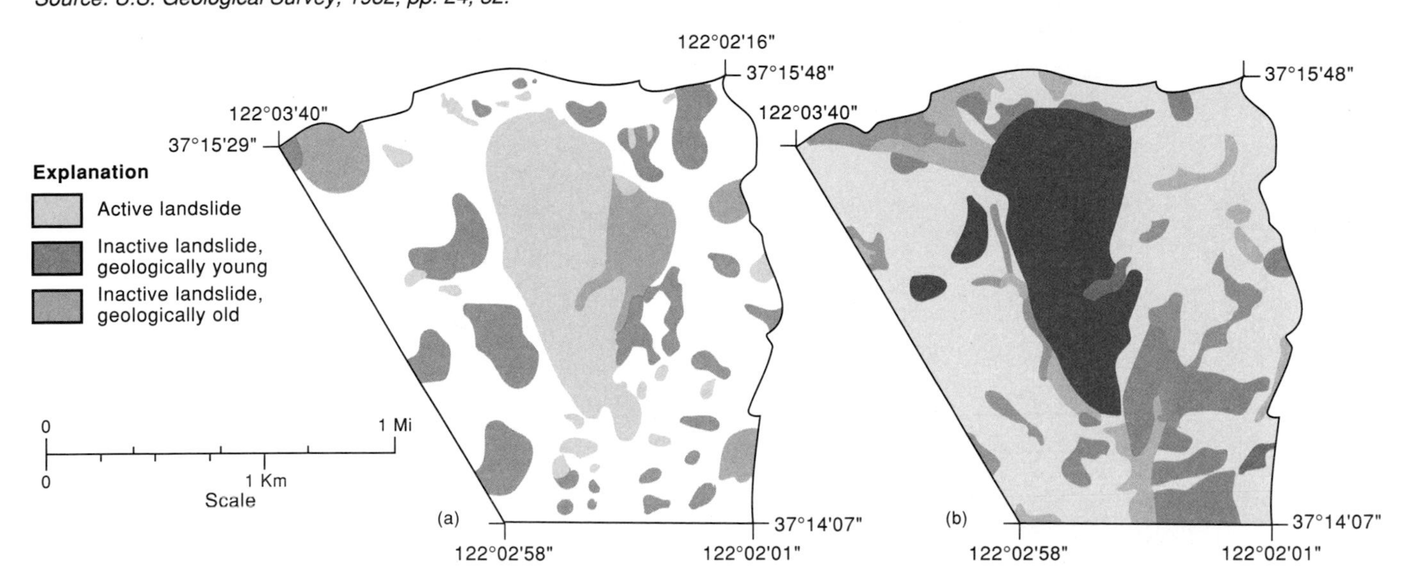


| Relative Stability | Map Area | Geologic Conditions | Recommended Land Use: Houses | Roads: Public | Roads: Private |
|---|---|---|---|---|---|
| Most Stable | | Flat or gentle slopes; subject to local shallow sliding, soil creep and settlement | Yes | Yes | Yes |
| | | Gentle to moderately steep slopes in older stabilized landslide debris; subject to settlement, soil creep, and shallow and deep landsliding | Yes* | Yes* | Yes* |
| | | Steep to very steep slopes; subject to mass-wasting by soil creep, slumping and rock fall | Yes* | Yes* | Yes* |
| | | Gentle to very steep slopes in unstable material subject to sliding, slumping, and soil creep | No* | No* | No* |
| | | Moving, shallow (<10 ft) landslide | No* | No* | No* |
| Least Stable | | Moving, deep landslide, subject to rapid failure | No | No | No |

Yes* - The land use would normally be permitted, provided the geologic data and/or engineering solutions are favorable. However, in some instances the use would be inappropriate.

No* - The land use would normally not be permitted. However, under some circumstances geologic data and/or engineering solutions would permit the use.


Matti, J. C., and Carson, S. E., 1991, *Liquefaction Susceptibility in the San Bernardino Valley and Vicinity, Southern California: A Regional Evaluation.* U.S. Geological Survey Bulletin 1898, 53 pp.

Mears, A. I., 1979, *Colorado Snow-Avalanche Area Studies and Guidelines for Avalanche-Hazard Planning.* Colorado Geological Survey Special Publication 7, 124 pp.

Nilsen, T. H., et al., 1979, *Relative Slope Stability and Land-use Planning in the San Francisco Bay Region, California.* U.S. Geological Survey Professional Paper 944, 96 pp.

Sadler, P. M., and Morton, D. M. (eds.), 1989, *Landslides in a Semi-arid Environment.* Riverside, California, Inland Geological Society, 386 pp.

Schultz, A. P., and Southworth, C. S. (eds.), 1987, *Landslides of Eastern North America.* U.S. Geological Survey Circular 1008, 43 pp.

Schuster, R. L., Varnes, D. J., and Fleming, R. W., 1981, Hazards from ground failures; *Landslides: In Facing Geologic and Hydrologic Hazards, Earth-Science Considerations,* pp. 54–65. U.S. Geological Survey Professional Paper 1240-B.

U.S. Geological Survey, 1982, *Goals and Tasks of the Landslide Part of a Ground-failure Hazards Reduction Program.* U.S. Geological Survey Circular 880, 48 pp.

Wold, R. L., Jr., and Jochim, C. L., 1989, *Landslide Loss Reduction: A Guide for State and Local Government Planning.* Colorado Geological Survey Special Publication 33, 50 pp.

## Figure 12.7

Geologic map of the area surrounding Hollidaysburg, Pennsylvania.

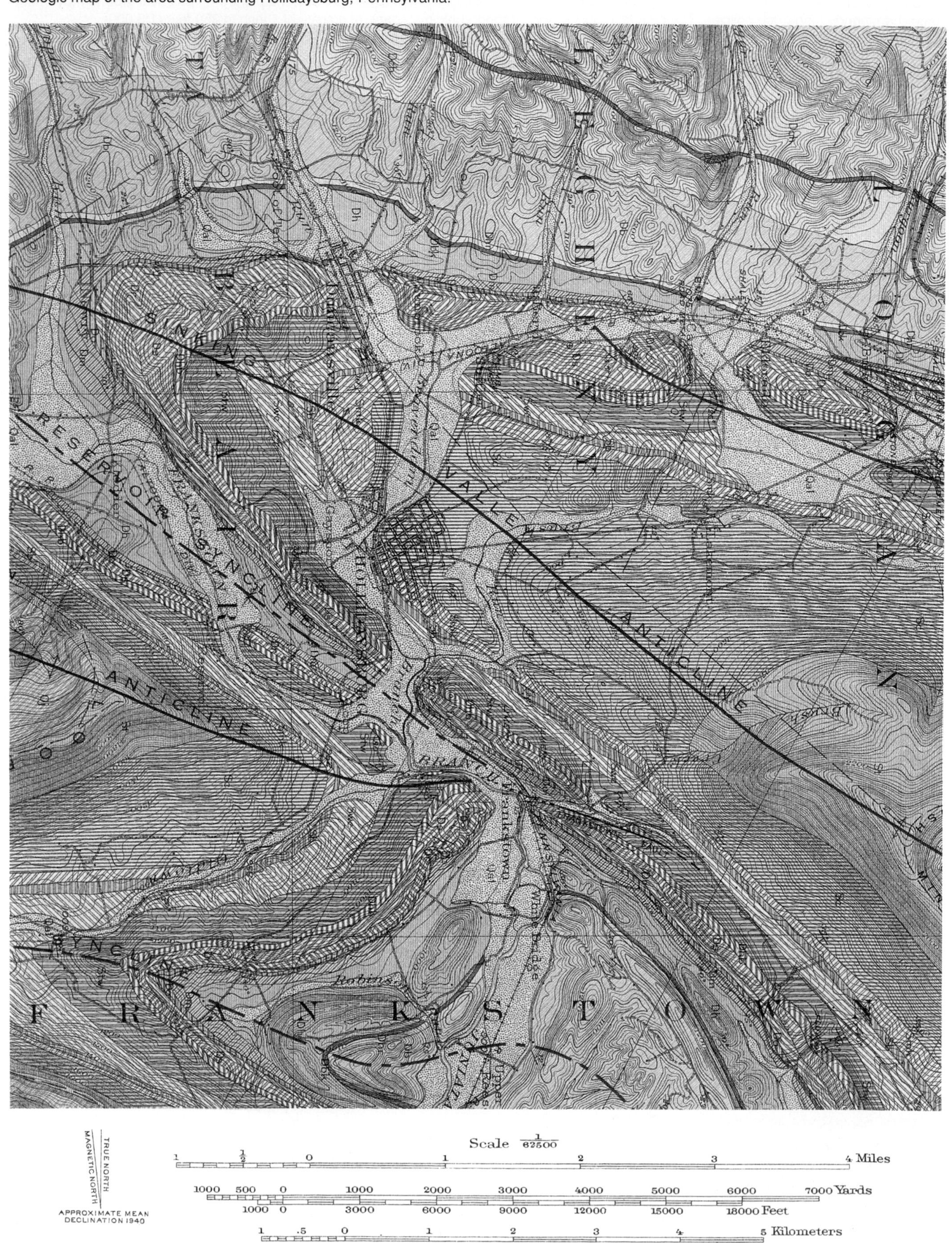

## Figure 12.7—*Continued*

EXPLANATION
SEDIMENTARY ROCKS

QUATERNARY

Recent

Qal
Alluvium
*(silt, sand, and gravel constituting the flood-plains of present streams)*

DEVONIAN

Upper Devonian

Dha
Hampshire formation
*(predominantly red, lumpy shale or mud rock and red sandstone; some gray and green shale and sandstone)*

Dsx Dch
Chemung formation
*(chiefly green, gray, and chocolate-colored shale and thin beds of argillaceous fine-grained sandstone; fossiliferous throughout; includes Saxton conglomerate member. Dsx: upper part largely chocolate-colored)*

Portage group

Db
Brallier shale
*(micaceous, siliceous slaty green shale with some thin beds of fine-grained sandstone; sparsely fossiliferous throughout, mainly pelecypods of Gardeau type)*

Dhr Dbk
Harrell shale
*(soft gray shale in upper part; Burket black shale member, Dbk, in lower part; highly fossiliferous, small pelecypods and cephalopods of the Naples fauna)*

Middle Devonian

Dh
Hamilton formation
*(principally olive-green shale with even-layered, blocky-jointed sandstone and thin limestone at top; ridge-making sandstone at two horizons; sparingly fossiliferous; locally a foot or two of limestone at top with Tully fauna)*

Dm
Marcellus shale
*(black fissile clay shale, grading upward into olive-green shale)*

Do
Onondaga formation
*(gray shale, probably calcareous, and thin argillaceous limestone)*

Lower Devonian

Oriskany group

Dr
Ridgeley sandstone
*(thick-bedded calcareous sandstone weathering to coarse friable sandstone; locally a fine conglomerate at top with quartz pebbles; highly fossiliferous)*

Ds
Shriver limestone
*(thin-bedded siliceous limestone, weathering to fine-grained sandstone; black calcareous shale at bottom; sparingly fossiliferous)*

Dhb
Helderberg limestone
*(lower part is thick-bedded gray limestone with thin gray chert at top, chiefly Keyser limestone member; overlying Coeymans and New Scotland limestone members thin and locally absent; contains valuable quarry rock, called "calico rock"; fossiliferous throughout)*

SILURIAN

Cayuga group

Stw
Tonoloway limestone
*(thin-bedded finely laminated, dark limestone; sparingly fossiliferous, chiefly Leperditia)*

Swc
Wills Creek shale
*(chiefly gray, calcareous shale and some greenish limestone; fossils scarce)*

Sb
Bloomsburg redbeds
*(lumpy red shale and thick-bedded ridge-making red sandstone)*

Smk
McKenzie formation
*(blue thin-bedded fossiliferous limestone and soft gray and green shale; thin red shale east of Tussey Mountain and a little red shale west of Lock Mountain)*

Niagara group

Sk Sc Scs
Clinton formation
*(mainly green and blue shale, weathering purplish, and thin fine-grained green sandstone in middle; Keefer sandstone member, Sk, near top; shale with thin limestone layers above Keefer sandstone member represent Rochester shale; Marklesburg iron-ore bed just beneath Keefer sandstone member; Frankstown iron-ore bed in lower half; hard quartzitic sandstone, red sandstone, and Levant Black iron ore, Scs at base; generally fossiliferous)*

St
Tuscarora quartzite
*(hard white quartzite and sandstone, largely thick-bedded; quartzite extensively quarried for ganister; contains scolithus worm tubes and Arthrophycus at top)*

ORDOVICIAN

Upper Ordovician

Oj
Juniata formation
*(chiefly red and some green fine-grained cross-bedded sandstone and red lumpy mud rock; nonfossiliferous)*

Oo
Oswego sandstone
*(gray fine-grained thick-bedded cross-laminated sandstone; contains a few small quartz pebbles in lower part; nonfossiliferous)*

exercise

# THIRTEEN

## Porosity, Permeability, and Fluid Flow through Rocks

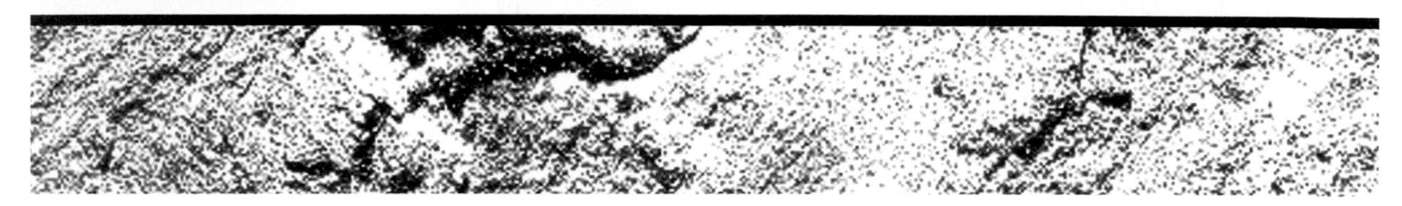

Not all precipitation is discharged at the surface as runoff; some percolates into soil, loose sediment, or rock and can descend many hundreds of feet into the ground. This groundwater is a common source of water for human consumption. The controls on the amount of infiltration are partly climatic (How much rain? How intense? For how long?), partly topographic (Are slopes steep or gentle? Is the rain ponded or free to flow downslope?), and partly lithologic. The lithologic controls involve the type of surface material, its porosity, and its permeability.

### *Porosity*

Porosity refers to the amount of void space between sediment particles:

$$\%\ \text{porosity} = \frac{\text{void volume}}{\text{total rock volume}} \times 100$$

$$\%\ \text{effective porosity} = \frac{\text{interconnected void volume}}{\text{total rock volume}} \times 100$$

A less used but still useful term is *void ratio.*

$$\text{void ratio} = \frac{\text{void volume}}{\text{grain volume}}$$

In general, only sedimentary materials have appreciable porosity. Igneous and metamorphic rocks are composed of interlocking crystals and, therefore, have no spaces between them that can hold water, although later fracturing can produce porosity in normally nonporous rocks such as basalt. In sandstones, interconnected porosity is usually about equal to or only slightly less than total porosity; limestones and dolostones, in contrast, can have many "dead-end" pores, reducing interconnected porosity to considerably less than total porosity.

The main controls of porosity in sedimentary materials are (1) grain size and sorting, (2) compaction, and (3) the amount of chemically precipitated cement among the clastic particles.

1. *Grain size and sorting.* Variations in mean grain size and sorting affect both the sizes of the pores in a clastic rock and the percentage of pore space. Coarser-grained sediments have larger pores, while better sorted sediments have larger pores and higher porosities. In poorly sorted sediments, which have a wider range of grain sizes, the smaller grains lodge in the pores among the larger grains, reducing both the percentage of pore space and the size of the pores.

   Further complicating porosity evaluation is the fact that fine-grained sediment can enter a clastic deposit after the grains are deposited. In a river setting, for example, the channel shifts location frequently, and floodplain mud commonly comes to lie above channel sand. The finer-grained mud then infiltrates downward into the pore spaces in the sand.

   Fine-grained soil particles formed on a sand can also infiltrate a sediment and decrease its porosity. Sediment can infiltrate a deposit only if the diameter of the infiltrated grains is less than one-tenth that of the host grains. For this reason only fine silt and clay can infiltrate into sand; sand can infiltrate only into gravel.
2. *Compaction.* As later sediments are deposited on top of earlier ones, the grains in the earlier ones are forced closer together, decreasing the size and

**Figure 13.1**

(a) Photomicrograph of quartz-cemented quartz sandstone with spaces (pores, dark gray areas) between grains made smaller by addition of quartz after burial of the sand. The view shows about 20 rounded quartz grains outlined by rings of dark material. Quartz was precipitated from groundwater on top of the dark rings, partially filling the depositional pore space. Clastic grains are about 0.5 mm in diameter. (b) Scanning electron micrograph of a sandstone showing incomplete filling of original pores. The flat surfaces on grains are small amounts of quartz precipitated from groundwaters. Many pores remain between grains, which are about 0.2 mm in diameter. Porosity is 13% and permeability is 127 millidarcies (md).

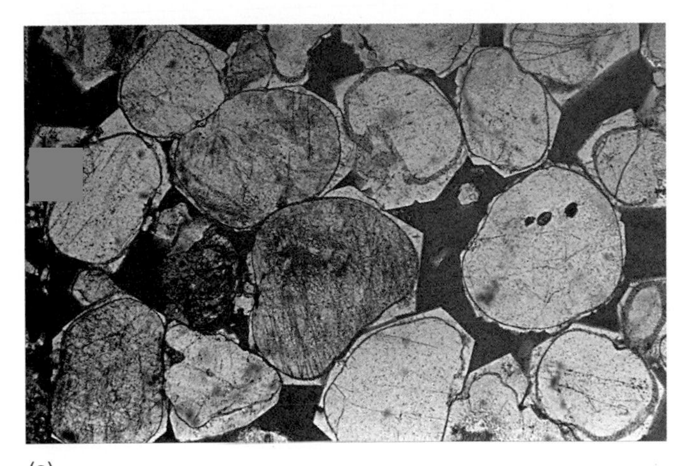

(a)

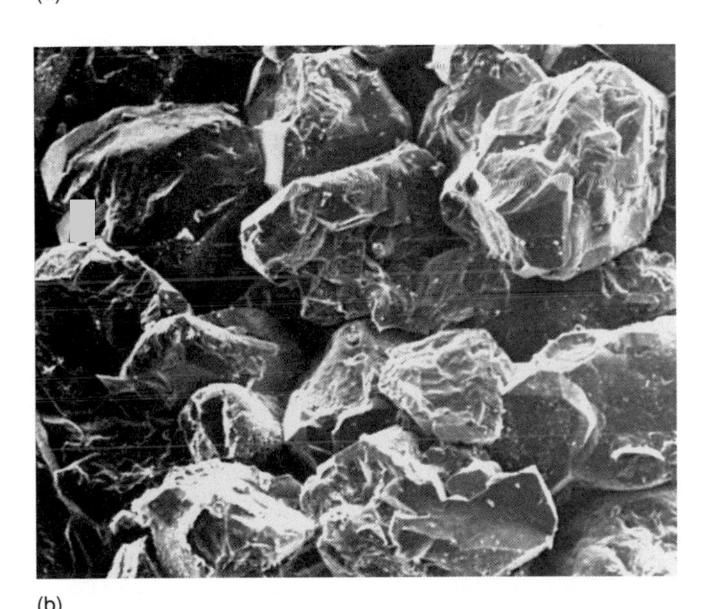

(b)

percentage of pores. A well-sorted quartz sand shows little decrease because quartz grains are hard and rigid. If, on the other hand, the sand contains a significant percentage of flexible and deformable fragments such as mica flakes, clay minerals, or pieces of schist, compaction is much more effective. Many sandstones had an estimated 30% pore space when deposited, but now have none because of burial, grain compaction, and squeezing of deformable fragments into depositional pores.

3. *Cementation.* From the time a layer of sediment is deposited, water percolates through it. This process continues to varying degrees for the life of the sediment, with the chemical composition of the water changing continually as it interacts with the mineral grains in the sediment. The sediment-water mixture is the site of an ongoing chemical reaction for many millions of years. The result is the precipitation of new minerals, most commonly calcite and quartz, from the pore water (figure 13.1). Thus, most ancient aggregates of sand grains (sandstones) are cemented by these two minerals. The only other common cement precipitate is hematite, which is easily identified by its red-brown color. Ancient sandstones typically have porosities of less than 10%.

Pore spaces in sedimentary rocks are located among clastic grains. As a result the spaces are small, normally smaller than the diameters of the grains in the rock. An average sandstone is formed of grains perhaps 0.3 mm (0.01 in) in size. Despite their small size, pores in sedimentary rocks can be numerous enough to hold a very large amount of fluid. Consider, for example, a sandstone layer 10 ft thick that extends over an area of 1 $mi^2$. The volume of the rock is 278,784,000 $ft^3$; if it contains 10% pore space, the volume of pore space is 27,878,400 $ft^3$. One gallon of water occupies 0.134 $ft^3$, so the pores in this layer of sandstone can hold more than 200 million gallons of water. The average American uses about 500 gallons per day for household purposes, so 200 million gallons of water would supply the household needs of a city of 400,000 for one day.

## *Hydraulic Conductivity and Permeability*

Hydraulic conductivity refers to the ease with which water flows through a rock and is defined by a relationship first recognized in 1856 by the French hydrologist Henri Darcy.

$$V = \frac{Q}{A} = K\frac{\Delta h}{\Delta l}$$

where $V$ = apparent velocity (cm/s)
$Q$ = discharge ($cm^3$/s)
$A$ = cross-sectional area ($cm^2$)
$K$ = hydraulic conductivity (cm/s)
$\Delta l$ = distance of flow (cm)
$\Delta h$ = hydraulic head (cm)

TABLE 13.1

Range of Values of Hydraulic Conductivity and Permeability.

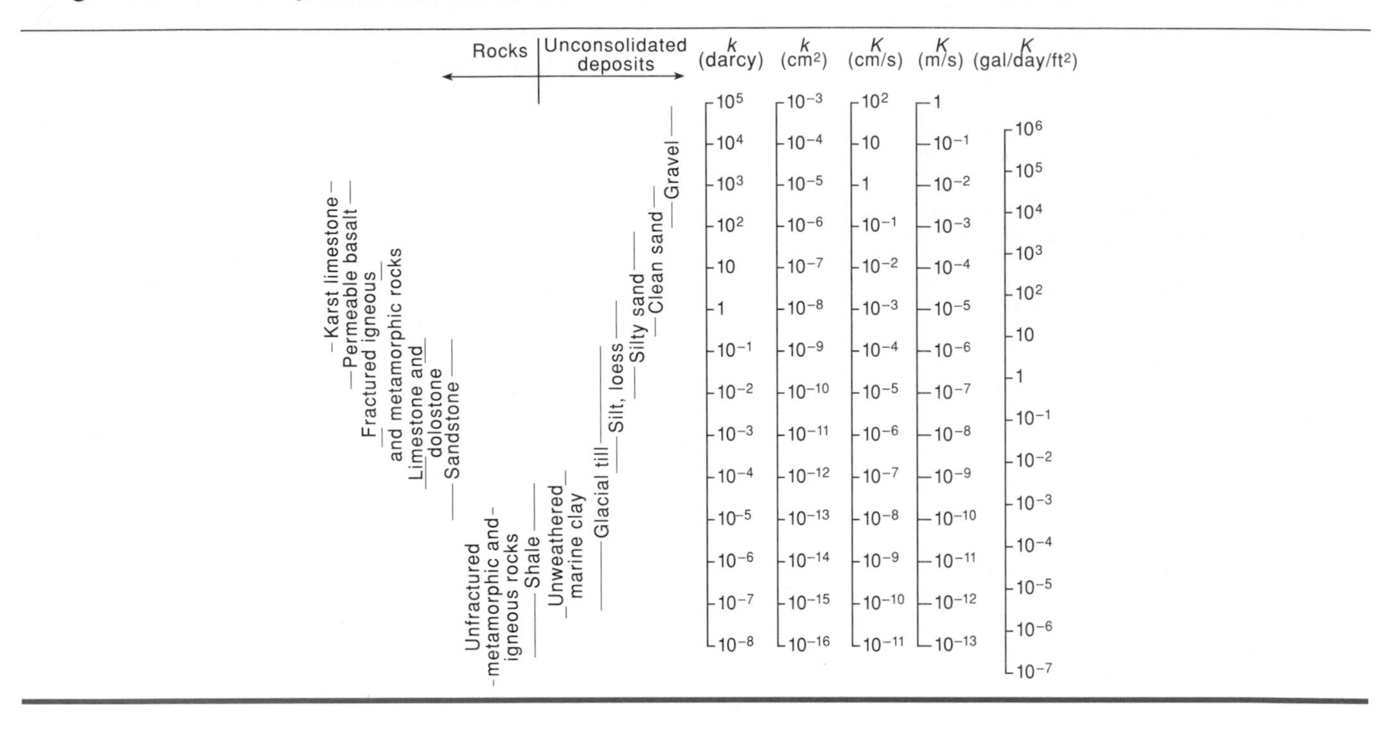


Hydraulic conductivity is related to *permeability* (a term used by petroleum geologists):

$$K = k\frac{\rho g}{\mu}$$

where $k$ = permeability (darcies = $cm^2 \times 10^{-8}$)
$\rho$ = fluid density ($gm/cm^3$)
$\mu$ = fluid viscosity (centipoises [cp], gm/cm s $\times 10^{-2}$)
$g$ = gravitational acceleration ($cm/s^2$)

Permeability is measured in units called darcies, after the French hydrologist. One darcy is the permeability that results in a discharge of 1 cm/s through 1 $cm^2$, for a fluid with a viscosity of 1 cp under a hydraulic gradient that makes the term ($\rho$, $g$ dh/dl) equal to 1 atm/cm. One darcy is approximately $10^{-8}$ $cm^2$. The permeabilities of oil-producing rocks are on the order of 0.1 darcies, or 100 millidarcies (md).

We can use these and related formulas to calculate the volume of available water in a water-bearing rock unit. For example, suppose a confined aquifer is 20 m thick and 8 km wide. Wells have been drilled 1.5 km apart in the direction of flow. The head in one well is 50 m and in the other is 46 m. The hydraulic conductivity is 1.4 m/day. How much water will flow through the aquifer each day?

$$Q = K\frac{\Delta h}{\Delta l}\text{(aquifer width) (aquifer thickness)}$$

$$Q = (1.4\text{ m/day})\frac{50\text{ m} - 46\text{ m}}{1{,}500\text{ m}}(20\text{ m})\ (8{,}000\text{ m}) = 597.3\text{ m}^3\text{/day}$$

Hydraulic conductivity and permeability in sedimentary rocks are normally greatest parallel to the layering (table 13.1) but can be greater perpendicular to the layering if the rock is highly fractured.

The chief lithologic controls on hydraulic conductivity are pore size (coarser-grained clastic rocks have larger pores) and the amount and type of clay minerals in the rock. Clay and smaller pores greatly increase the amount of surface over which a fluid must flow in its path through the rock, and the increased area of "friction" (actually electrostatic attraction) between the fluid and the grain surfaces reduces the flow velocity of the fluid. Clay minerals are shaped like sheets of paper, a shape with much more surface area than that of a more equidimensional particle, such as a quartz grain, with the same volume. A few percent clay in a sandstone is enough to virtually eliminate the flow of water or petroleum through an otherwise productive layer.

Mudrocks cannot transmit fluids unless fractured. This lack of fluid conductivity results from the fact that the average mudrock contains about 60% easily flexible clay flakes, grains that bend and conform to one another's shape when compacted (figure 13.2).

Limestones are composed almost entirely of calcium carbonate, which is fairly easily dissolved (soluble) in underground waters (figure 13.3). Sometimes the dissolution extends through a very large volume, producing a cavern of considerable size. Carlsbad Cavern in New Mexico and Mammoth Cave in Kentucky are examples of huge dissolution cavities in limestones (figure 13.4).

**Figure 13.2**

Scanning electron micrograph of a shale showing the parallelism of clay flakes caused by the compaction that accompanies burial of the original shallow-sea mud. This rock has no permeability to fluids.

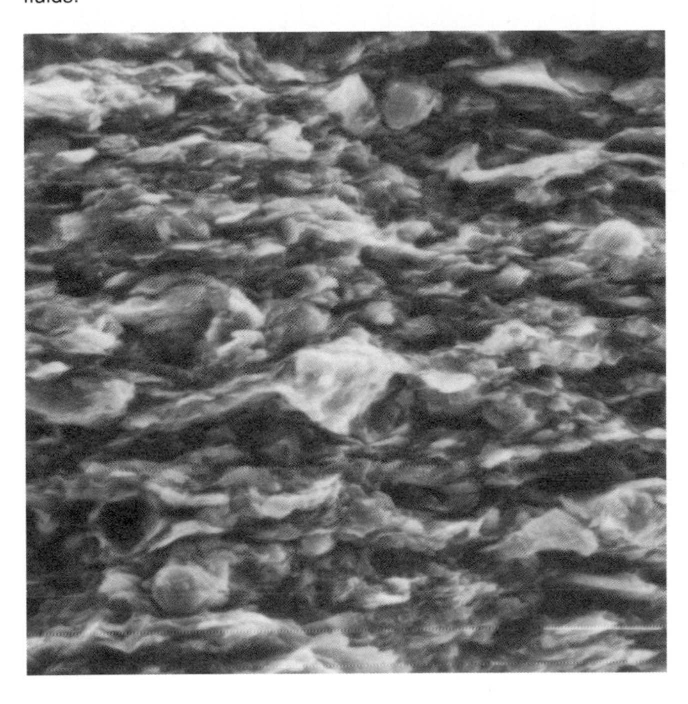

**Figure 13.3**

Partially interconnected (in the third dimension) solution cavities in a drill core of microcrystalline (aphanitic) limestone. Width of core is 2.5 inches.

## *Porosity, Permeability, and Depositional Environment*

Grain size, degree of sorting, and clay content in a sediment depend mostly on the level of kinetic energy in the depositional environment. We can see this relationship in modern environments. A beach setting, for example, has a very high level of kinetic energy because of the pounding surf and wave movement; so does a sand-dune environment because of its strong winds. As a result, the sand and gravel in these environments are well-sorted and contain no clay, which is winnowed out by moving water or air. In contrast, levee and floodplain sands along river margins are poorly sorted and typically contain much clay. Better-sorted and clay-free sands have higher porosities and hydraulic conductivities than those of sands with less mature textures; therefore, most of the important, areally extensive, water-bearing subsurface rocks are ancient beach sands, nearshore barrier bars, stream-channel sands, glacial outwash sands, and desert dunes.

## Problems

1. In a confined aquifer, hydraulic conductivity is 7.1 m/day, and the hydraulic head in the direction of flow differs by 6.2 m in two test wells drilled 1.7 km apart. What is the apparent velocity of water flow between the two wells?
2. Suppose you wished to increase the flow rate of underground water in a low-permeability, calcite-cemented sandstone. How might you try? Would your method work equally well in a limestone? Why or why not?
3. The rock property most important to permeability is internal surface area. Using log-log graph paper, show the relationship between the diameter of spherical quartz grains and their surface-to-volume ratio. (The surface area of a sphere is $\pi d^2$; the volume of a sphere is $1/6\ \pi d^3$, where $d$ = grain diameter.)
4. An average clay flake has the dimensions 1 μm × 1 μm × 0.01 μm. What are the volume and surface area of this clastic grain? What is the diameter of a quartz or feldspar sphere with the same diameter as the length of

**Figure 13.4**

Extremely large "pores" in limestone, Carlsbad Cavern, New Mexico. The pore in which the man is standing bifurcates into two smaller ones in the direction he is facing. Limestone caverns are simply very large pore networks formed by dissolution of soluble calcium carbonate.

a clay flake? What is the volume of this sphere? What do these results tell you about the relative importance of quartz and clay in controlling permeability?

5. Some rocks have significant porosity (10–20%) but low (or no) permeability. How is this possible? Describe the character and internal texture of at least two different kinds of rocks that might fit this description.
6. Figure 13.5 is a graph showing changes in water density and viscosity with changing temperature. Using these data, graph the effect on hydraulic conductivity for a sandstone with a permeability of $10^{-3}$ darcies as temperature increases from 20°C to 200°C, a temperature reached at an average depth of about 23,000 feet.
7. As an environmental geologist you are asked to design at least one laboratory experiment to determine quantitatively the effect of grain size on permeability to water. Not wanting to miss out on a fat consulting fee, you agree to try. Suggest possible approaches. Be as specific as you can.

**Figure 13.5**

Change in the viscosity and density of water as a function of temperature.

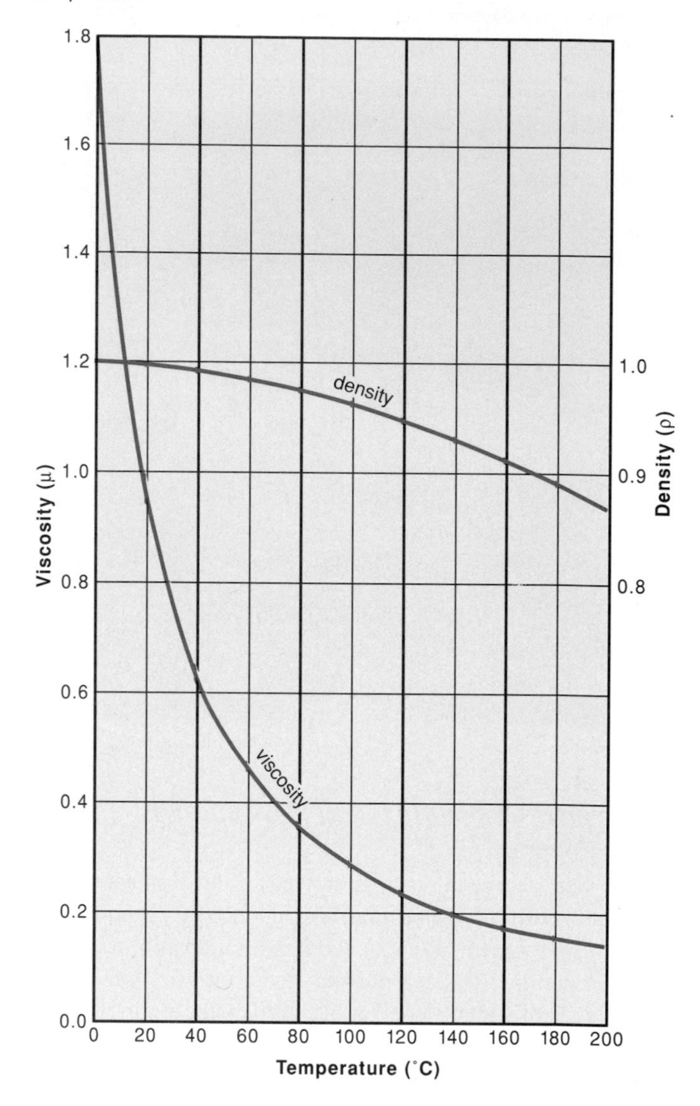


## Further Reading/References


Anderson, G., 1975, *Coring and Core Analysis Handbook.* Tulsa, Oklahoma, Petroleum Publishing Co., 200 pp.

Domenico, P. A., and Schwartz, F. W., 1990, *Physical and Chemical Hydrogeology.* New York, John Wiley & Sons, 824 pp.

Fetter, C. W., 1988 *Applied Hydrogeology.* 2nd ed., New York, Macmillan, 592 pp.

Phillips, O. M., 1991, *Flow and Reactions in Permeable Rocks.* New York, Cambridge University Press, 285 pp.

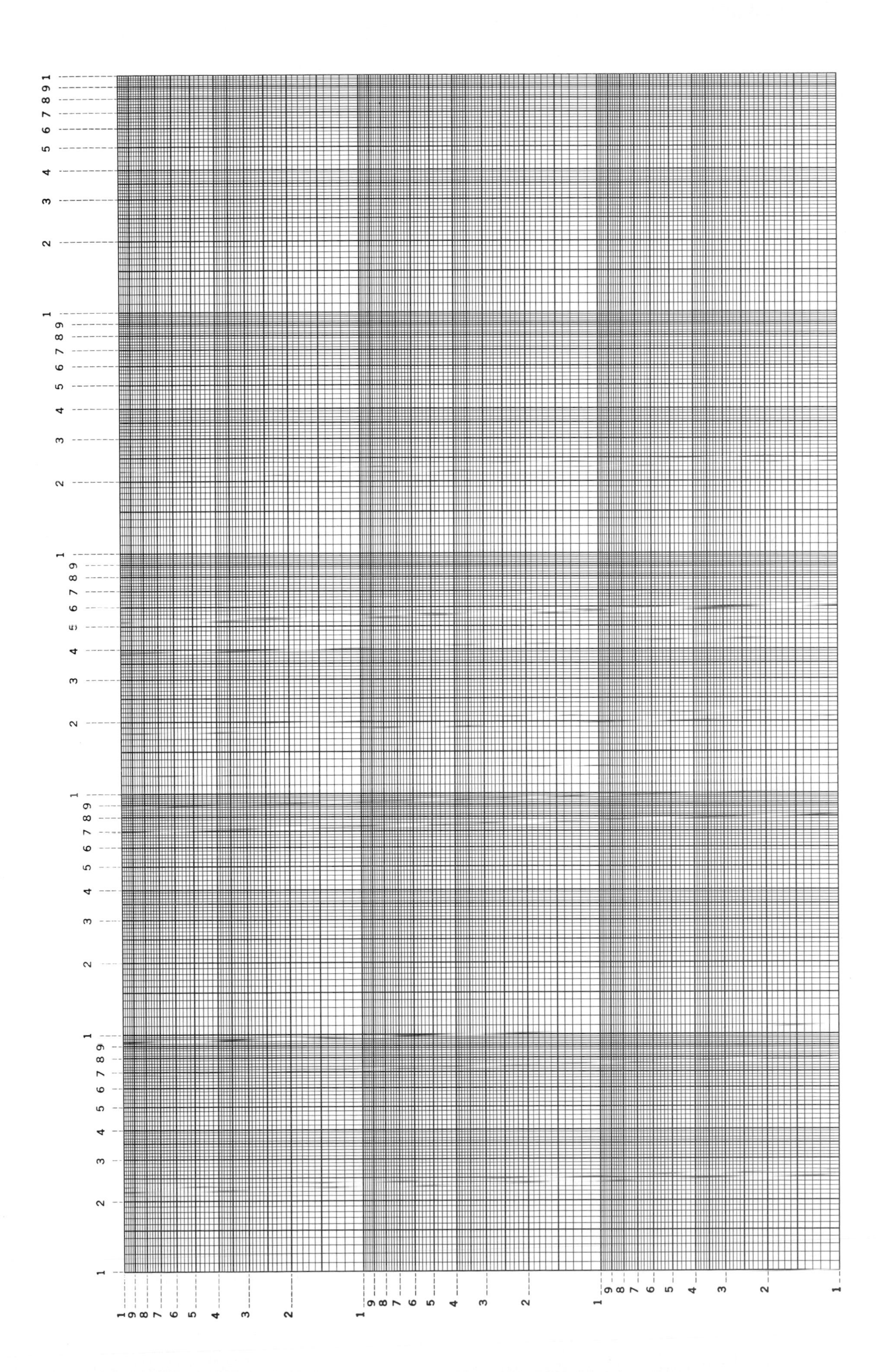

exercise

# FOURTEEN

## *Groundwater*

Groundwater is one of our most important natural resources, currently supplying 20–25% of the water used in the United States. Subsurface water can occur in any porous sediment or rock, but amounts large enough for intensive exploitation are found mostly in sedimentary materials. These materials underlie about 65% of the world's land surface, making groundwater a widely distributed natural resource. In most areas of sedimentary rocks it is easier to drill a well and find water than to drill and not find it. This fact explains most of the successes attributed to water witchers. Drilling in areas of igneous and metamorphic rocks, however, produces groundwater only in some cases in which the rocks are fractured. Lava flows commonly crack and fragment at their upper surfaces because of rapid cooling; when they are buried beneath younger materials, they retain pores and can become important water sources, as they are in several northwestern states and Hawaii.

Sediments and rocks that yield water in amounts large enough to be significant to humans are called *aquifers* (figure 14.1). Some aquifers are unconfined, meaning that the water-bearing unit extends up to the ground surface; others are capped by an impermeable confining layer, called an *aquiclude* or *aquitard,* that does not transmit fluid. Confined aquifers occur at depth, unconfined aquifers near the ground surface. Any aquifer, whether confined or unconfined, has a level below which the pores are full of water and above which they contain mostly air mixed with small amounts of water held between adjacent grains by capillary forces. The surface or narrow zone between these two regions is called the *water table.* Wells drilled for water production must penetrate to beneath the water table.

Environmental geologists are interested in (1) depth to the water table; (2) direction and velocity of flow and discharge of the aquifer; (3) replenishment rate and storage capacity of the aquifer; and (4) the sources, types, and movement of possible pollutants.

Water enters the ground from precipitation, lakes, and streams, filtering downward through permeable materials until it reaches the water table. The water table can occur at any level between the basal aquiclude and the ground surface. If a subsurface water table intersects a permeable fracture that extends upward to the ground surface, a spring forms (figures 14.2, 14.3). Springs also form where the water table intersects a hillside, making them common in mountainous areas. Contrary to advertising claims, this spring water is no more or less healthful than unpolluted water from other sources. The level of the water table is determined by the balance between inflow rate to the aquifer and the rate of withdrawal or discharge. Generally an unconfined water table has an irregular surface shaped like a subdued replica of the overlying topography. The high areas of the water table are groundwater divides, the equivalent of above-ground topographic divides. Both

**Figure 14.1**

Cross-section showing confined and unconfined aquifers. The potentiometric surface is the level to which the groundwater rises without being pumped. The cone of depression reflects a rate of pumping greater than the water's rate of replenishment.

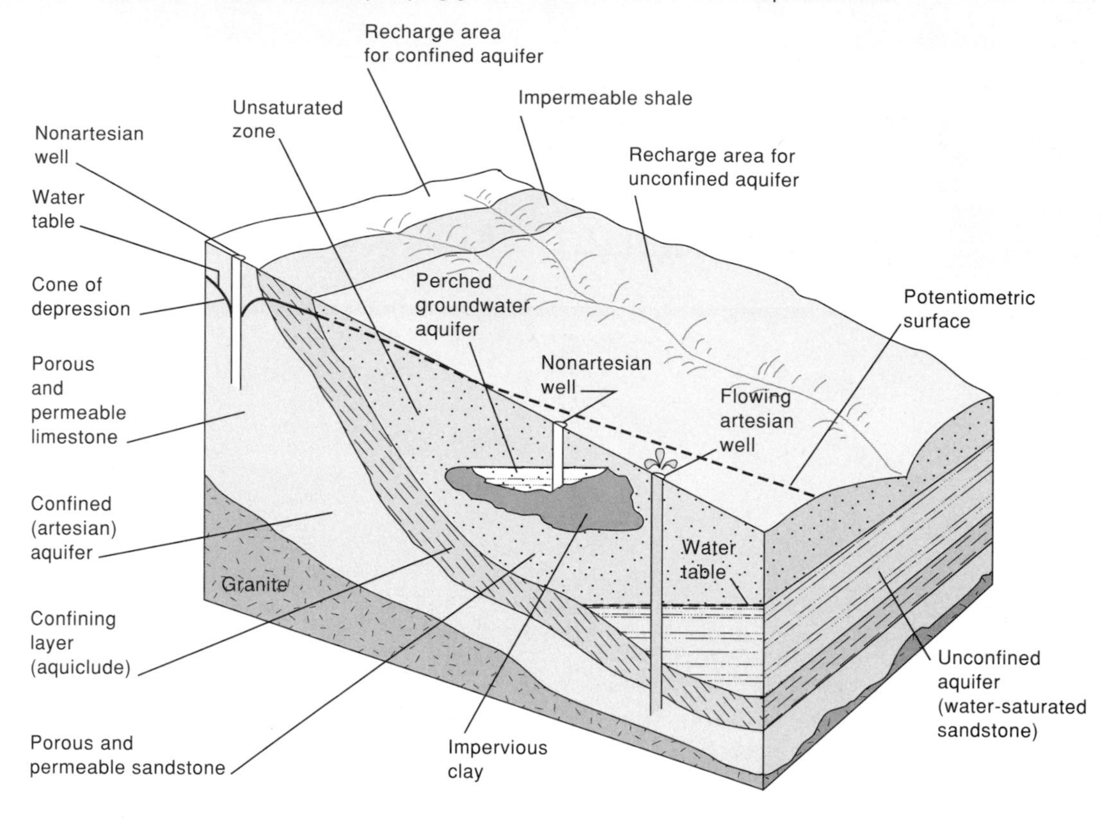


above and below ground, water flows away from the divides. Drier climates have deeper groundwater tables and less pronounced similarity between surface topography and the water table.

## *Groundwater Movement*

The direction of water movement beneath the ground surface depends on several variables. Water entering an unsaturated, homogeneous body of permeable sediment, such as some glacial deposits or alluvial valley fills, moves vertically downward because of gravity. Deviations from the vertical can occur, however, if permeability within the sediment varies. Water follows the path of least resistance and moves most readily through parts of the sediment that have the highest permeability.

Velocity of flow below the water table varies greatly among different aquifers but is typically a few inches per year—much slower than the velocity in the overlying unsaturated zone—and is described by a form of Darcy's law:

$$V = \frac{K}{n_e} \frac{\Delta h}{\Delta l}$$

where $n_e$ is the decimal percentage of interconnected pore space, $v$ is linear velocity (in cm/s), $\Delta h$ is the difference (in cm) in head between two points in the path of movement; $\Delta l$ is the distance (in cm) along the path of movement, and $K$ is hydraulic conductivity (in cm/s). The difference in head between two points ($\Delta h$) is termed the total head and is the height of the water above an arbitrarily chosen datum plane; it is composed of an elevation head and a pressure head. The term $\Delta h / \Delta l$ is the hydraulic gradient. The distribution of total head over an area defines the *potentiometric (piezometric) surface*. This surface is the level to which the water would rise if the aquiclude were punctured.

Permeability in the Darcy equation is a variable whose value depends only on properties of the rock, not on properties of the fluid. The term is used regularly in the petroleum industry, where it was first applied to commercial endeavors. In groundwater research, the term hydraulic conductivity is used more commonly than is permeability. Hydraulic conductivity is the rate of flow of a unit volume of water per day under a unit hydraulic gradient through a unit cross-sectional area at prevailing temperatures.

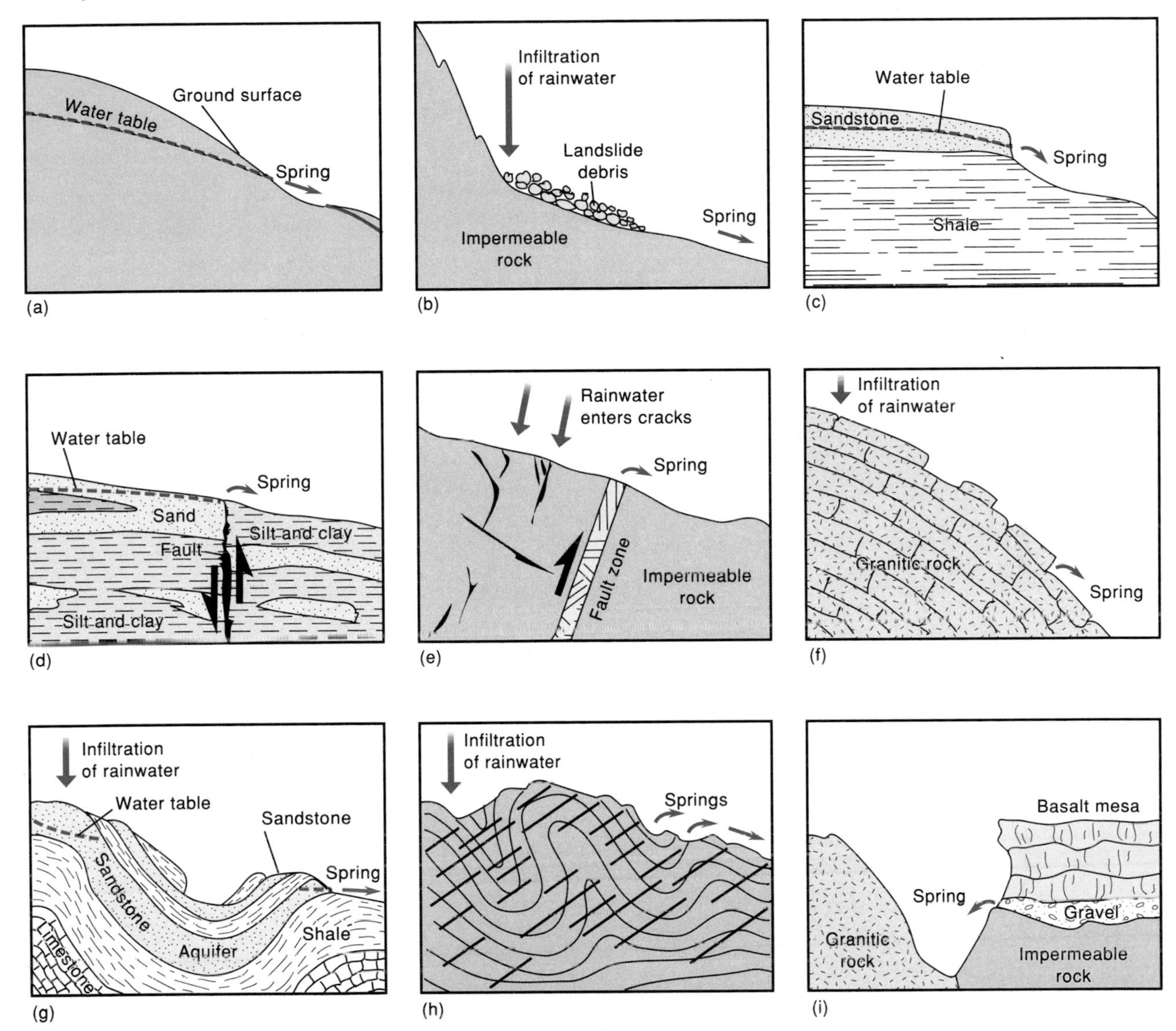


**Figure 14.2**

Geological factors in the location of springs.

*From S. N. Davis and R. J. M. DeWiest,* Hydrogeology. *Copyright © 1966 by John Wiley & Sons, Inc., New York, NY. Reprinted by permission of John Wiley & Sons, Inc.*

Hydraulic conductivity is thus a term specifically designed for use in studying subsurface water and cannot be used for other fluids, such as petroleum or natural gas.

Hydraulic conductivity is related to discharge, flow velocity, porosity, and potentiometric gradient by the following equations:

$Q = VA$ and $V = KI$, so that $Q = KIA$

and $v = KI/n_e$

$Q$ = discharge ($ft^3$/day)
$V$ = Darcy velocity of flow (ft /day)
$v$ = average linear velocity of flow (ft /day)
$A$ = cross-sectional area of the aquifer ($ft^2$)
$K$ = hydraulic conductivity (ft /day)
$I$ = change in head divided by the distance travelled; slope of the water table; potentiometric gradient (ft /ft)

Typically, water wells drilled into confined aquifers are drilled where the rock is deeper than the recharge site, which is usually the ground surface. The groundwater at depth is under pressure from the water updip in the aquifer. As a result, when the drill pierces the aquifer, the water rises above the base of the confining bed. A well in which this rise occurs is an *artesian well.* Sometimes the pressure is great enough to make the water rise above ground level, but in most wells the water rises only part way to the surface and must be pumped the rest of the way. Free-flowing water wells were more common in the past (figure 14.4), before intensive subsurface-water use lowered regional

**Figure 14.3**

Field example of a natural spring in the setting shown in figure 14.2c. Note that there is no stream behind the two brothers. The water issues from the rock immediately in front of them.

**Figure 14.4**

Photograph of one of the highest artesian well flows on record, drilled in 1909 into the Dakota Formation (Cretaceous) at a depth of about 775 feet in Woonsocket, South Dakota. The well was 6 inches in diameter and had a flow of 1,150 gallons per minute, and the water spout rose more than 100 feet above the ground surface.

water levels and reduced the pressure head in confined aquifers. Water yields from aquifers range from a few gallons per hour to perhaps 20 gallons per second in very permeable rocks with high hydraulic gradients.

Several different types of maps are used to show variations in groundwater movement over an area. Contour maps can show variables such as hydraulic gradient, water-table height, potentiometric surface, flow velocity, or changes in water-table height as water is withdrawn from an aquifer. Such maps can be compared to maps of aquifer thickness (isopach maps), lithologic characteristics, or permeability. Each type of map supplies a different kind of insight into the variables that control the availability of subsurface water.

## *Saltwater Encroachment*

Most of the sedimentary rocks in the geologic column were deposited in the oceans, so their original pore water was sea water. Subsequent flushing by surface water often has caused freshening at shallow depths, so that salinity of pore waters generally increases with depth. Potable waters

**Figure 14.5**

(a) A freshwater lens as it exists on an island composed of a uniform and homogeneous sandstone: $h_s$ is the depth below sea level of the interface between fresh water and salt water; $h_f$ is the height of the water table above sea level; $h_s/h_f = 40$. (b) Encroachment of salt water into the lens of fresh water as a result of pumping and removal of fresh water near a shoreline.

*Source: R. C. Heath, 1983, U.S. Geological Survey* Water-Supply Paper 2220.

Freshwater lens floating on salt water

Land surface

$h_f$

Water table

Sea

Sea

Fresh water $h_s$

Salt water

Zone of diffusion

(a)

Two aspects of saltwater encroachment

Pumping well

Land surface

Water table

Sea

Unconfined aquifer

Fresh water

Confining bed

Confined aquifer

Lateral encroachment

Upconing of salty water

Salt water

Confining bed

(b)

**Figure 14.6**

Chloride concentrations in water samples from the Union Beach Borough well field, 1950–1977.

*Source: F. L. Schaefer and R. L. Walker, 1981, U.S. Geological Survey* Water-Supply Paper 2184.

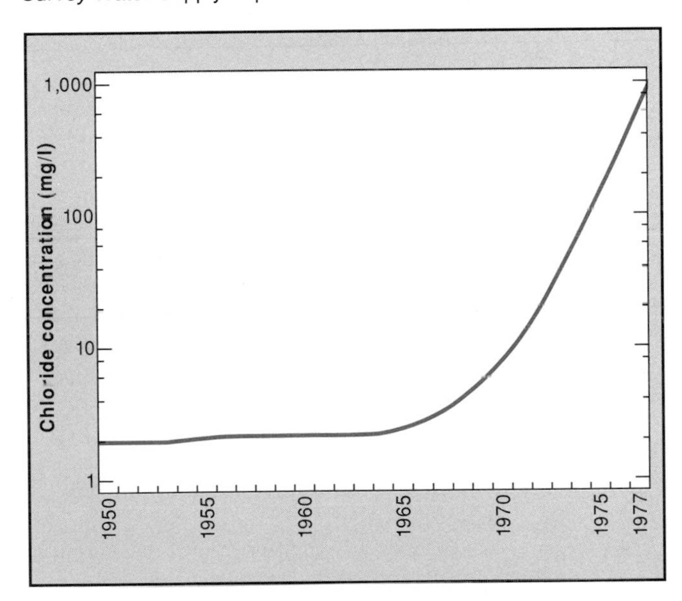


(those safe and palatable for human use) can occur to depths of perhaps 1 km, but most water wells are limited to depths of about 100 m because of drilling costs.

The principles by which water occurs in and moves through porous media apply to saline as well as fresh water. Encroachment of salt water into a freshwater aquifer can be induced in many ways, such as by pumping too much fresh water from wells; by puncturing the aquitards that protect freshwater aquifers (by wells, tunnels, dredging, or other construction); by ponding or otherwise enabling saline water to move downward into fresh water aquifers; or by discharging saline wastes directly into aquifers. Salty water, including sea water, is denser than fresh water and thus tends to move inland under it, creating a freshwater-saltwater interface. This interface occurs at a depth below sea level of about 40 times the height of the freshwater table above sea level (figure 14.5). This 40/1 relationship expresses a hydrostatic equilibrium between two fluids of different densities (sea water has a density of 1.025 gm/cm$^3$).

An example of saltwater encroachment is found in the case of water wells in Union Beach, New Jersey (figure 14.6). Until the early 1960s the chloride-ion concentration, a commonly used measure of the presence of sea water, was static at about 2 mg/l despite lowering of the water table by 15 ft caused by withdrawals from wells. But as water use continued to increase and the water table continued to drop below sea level, the hydraulic head of the fresh water became inadequate to keep the sea water from infiltrating the aquifer. By 1977 the chloride concentration reached 660 mg/l (sea water contains 19,000 mg/l chloride ion). Chloride concentrations greater than 250 mg/l exceed the public health standards for potable water.

## *Groundwater and Karst*

Although groundwater can move through sandy sediments at rates of several feet per day, most moves through the pore spaces of rocks very slowly. A typical velocity might be a few inches per year. Such a slow-moving fluid cannot cause mechanical erosion on the pore walls, but it can cause chemical erosion by dissolving the pore walls (the grains of the rock) if it is undersaturated with respect to the mineral forming the walls. Such is the case with shallow groundwaters and calcite. Fresh water percolating downward through soils becomes enriched in $CO_2$ from the decomposition of soil organic matter, and as the water enters underlying limestone this high $CO_2$ level makes it undersaturated with calcite. Over thousands of years the acidity (pH ~5) can dissolve large holes. When the holes are large enough for humans to tour, they are called caverns. Well-known examples are Carlsbad Cavern in New Mexico and Mammoth Cave in Kentucky. The largest room at Carlsbad is more than 4,000 ft long, 600 ft wide, and 350 ft high—a volume of nearly one billion cubic feet, testifying to the chemical aggressiveness of shallow groundwater in limestone. The resultant water is very "hard" (>75 ppm $Ca^{+2}$ + $Mg^{+2}$) and is not of the best quality for washing clothing or people.

The creation of large holes within a few hundred feet of the ground surface typically causes overlying rock layers to collapse, particularly when the rock contains vertical fractures (joints), as most limestones do. Such a roof-collapse is disastrous for people who live nearby (figure 14.7).

**Figure 14.7**

Aerial view of a large sinkhole that formed in Winter Park, Florida, in May, 1981. Several buildings have partially collapsed into the hole, which may enlarge further in the future.

## Problems

1. Suppose the water table is at the ground surface in a topographically flat area. What would be the result, that is, how would you recognize such an occurrence?
2. Suppose a stream is located above an unconfined aquifer. What will be the effect on streamflow?
3. Farmer Smith is having a water-well drilled. His instructions to the drillers are to save money by stopping as soon as they encounter water. Is this a smart decision? Explain.
4. Nassau and Suffolk counties are densely populated areas on Long Island, New York. Shown are a generalized geologic cross-section of the Long Island aquifer system in Nassau County (figure 14.8A) and the elevation of the water table on Long Island (figure 14.8B).
   a. How many different aquifers are present on Long Island? Name them.
   b. Which aquifers are confined? Unconfined?
   c. Is the freshwater supply in any of the aquifers interconnected?
   d. In which of the aquifers do you think saltwater intrusion might be a serious problem? Explain.
   e. Which aquifer would be most vulnerable to pollution from human activities?
   f. Draw a line to locate the groundwater divide on Long Island (figure 14.8B). Locate the divide on cross-section 14.8A. In which aquifer is the divide most effective?
   g. The thickness of the saturated zone shows a consistent increase from north to south on Long Island, to a maximum of about 1,000 ft at the south shore. Why does this increase occur? (Hint: examine figure 14.8A).
   h. Which aquifer should suffer most from saltwater intrusion along the south shore of Long Island?
   i. The flow velocity of water in the Magothy aquifer varies irregularly on Long Island. What can you infer about the lithologic character of this aquifer?
5. In the map on page 115, black lines show topography, while dashed lines signify elevation above sea level of the potentiometric surface of a confined aquifer.
   a. Outline the areas where wells will flow at the land surface without being pumped.
   b. Locate the best spot for drilling a water well, based on the relationship between topography and potentiometric surface.
   c. How far above the ground surface will the water spout at your well site?
   d. Locate the spot where a well would require the most pumping. How high would the water need to be pumped?
   e. In which direction is the recharge area for this aquifer?
   f. The potentiometric contours decrease uniformly from NE to SW. Why do you think these elevations decrease? What does this pattern tell you about the uniformity of the rock orientation of the aquifer layer in the area?
   g. Suppose you were going to purchase one-quarter of this map area on the basis of groundwater supply. Which would you choose: NW, NE, SE, or SW? Why?
6. What volume of water would flow through a valley filled with porous, permeable quartz sand 100 ft thick and 1 mi wide, where the hydraulic conductivity is

**Figure 14.8a–b**

(a) Generalized geologic section of the Long Island aquifer system in Nassau County. (b) Water-table altitude in Nassau and Suffolk Counties. *Source: U.S. Geological Survey Water-Resources Investigations Report 86-4141 and* Professional Paper 627-E, 800 C.

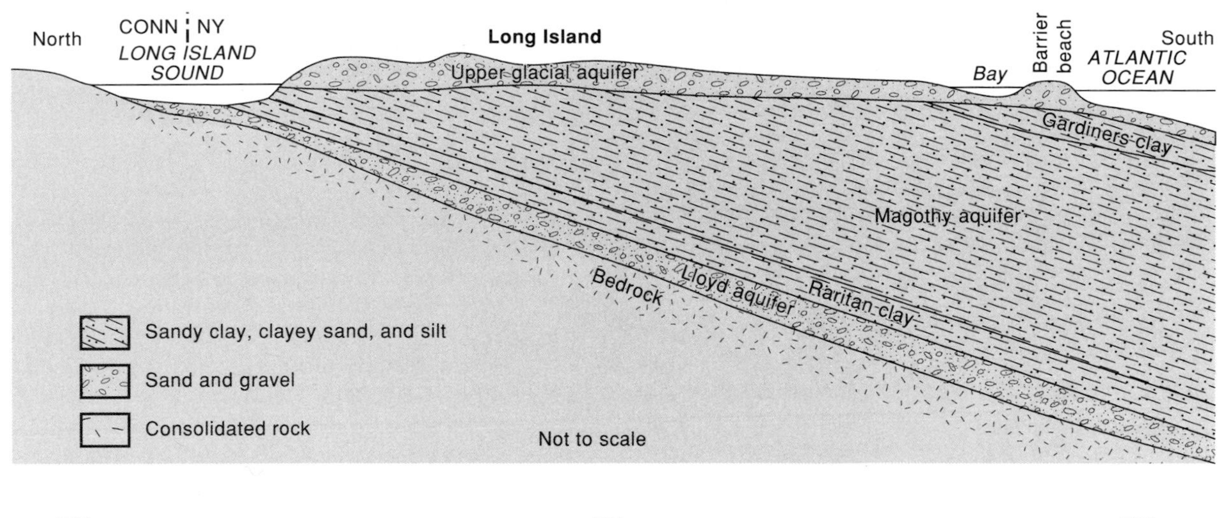


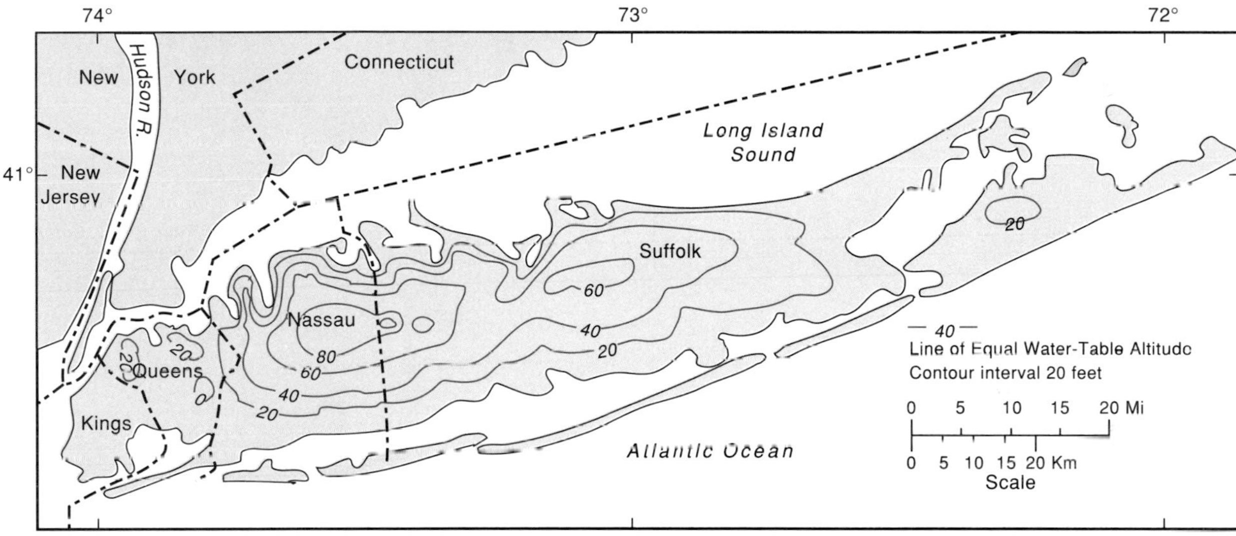


**Map for problem 5.**

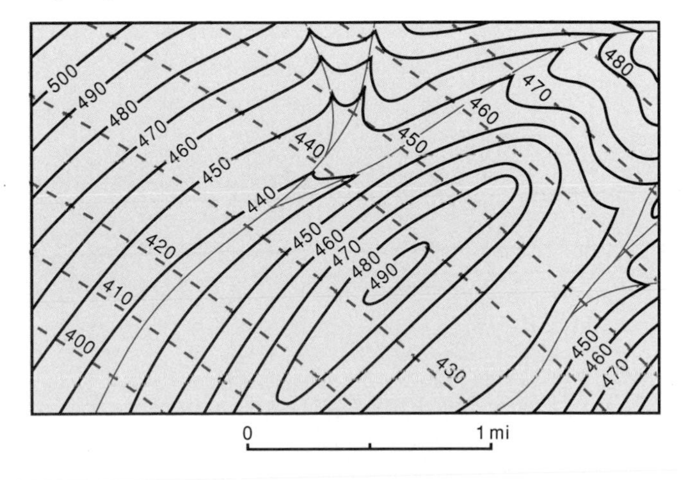

500 ft/day and the potentiometric gradient is 5 ft/mi? What is the linear flow velocity if the effective porosity is 20%?

## Further Reading/References


American Society of Civil Engineers, 1987, *Ground Water Management.* New York, 263 pp.

Beck, B. F. (ed.), 1989, *Engineering and Environmental Impacts of Sinkholes and Karst.* Rotterdam, A. A. Balkema, 384 pp.

Heath, R. C., 1983, *Basic Ground-water Hydrology.* U.S. Geological Survey Water-Supply Paper 2220, 84 pp.

Heath, R. C., 1989, *Ground-water Regions of the United States.* U.S. Geological Survey Water-Supply Paper 2242, 78 pp.

Higgins, C. G., and Coates, D. R. (eds.), 1990, *Groundwater Geomorphology.* Geological Society of America Special Paper 252, 368 pp.

Jorgensen, D. G. (ed.), 1984, *Geohydrology of the Dakota Aquifer.* Worthington, Ohio, National Water Well Association, 247 pp.

Newton, J. G., 1987, *Development of Sinkholes Resulting from Man's Activities in the Eastern United States.* U.S. Geological Survey Circular 968, 54 pp.

Palmer, A. N., 1991, Origin and morphology of limestone caves: *Geological Society of America Bulletin,* v. 103, pp. 1–21.

Schaefer, F. L., and Walker, R. L., 1981, *Saltwater Intrusion into the Old Bridge Aquifer in the Keyport-Union Beach Area of Monmouth County, New Jersey.* U.S. Geological Survey Water-Supply Paper 2184, 21 pp.

Todd, D. K., 1983, *Ground-water Resources of the United States.* Berkeley, California, Premier Press, 749 pp.

Troester, J. W., and Moore, J. E., 1989, Karst hydrogeology in the United States of America: *Episodes,* v. 12, no. 3, pp. 172–178.

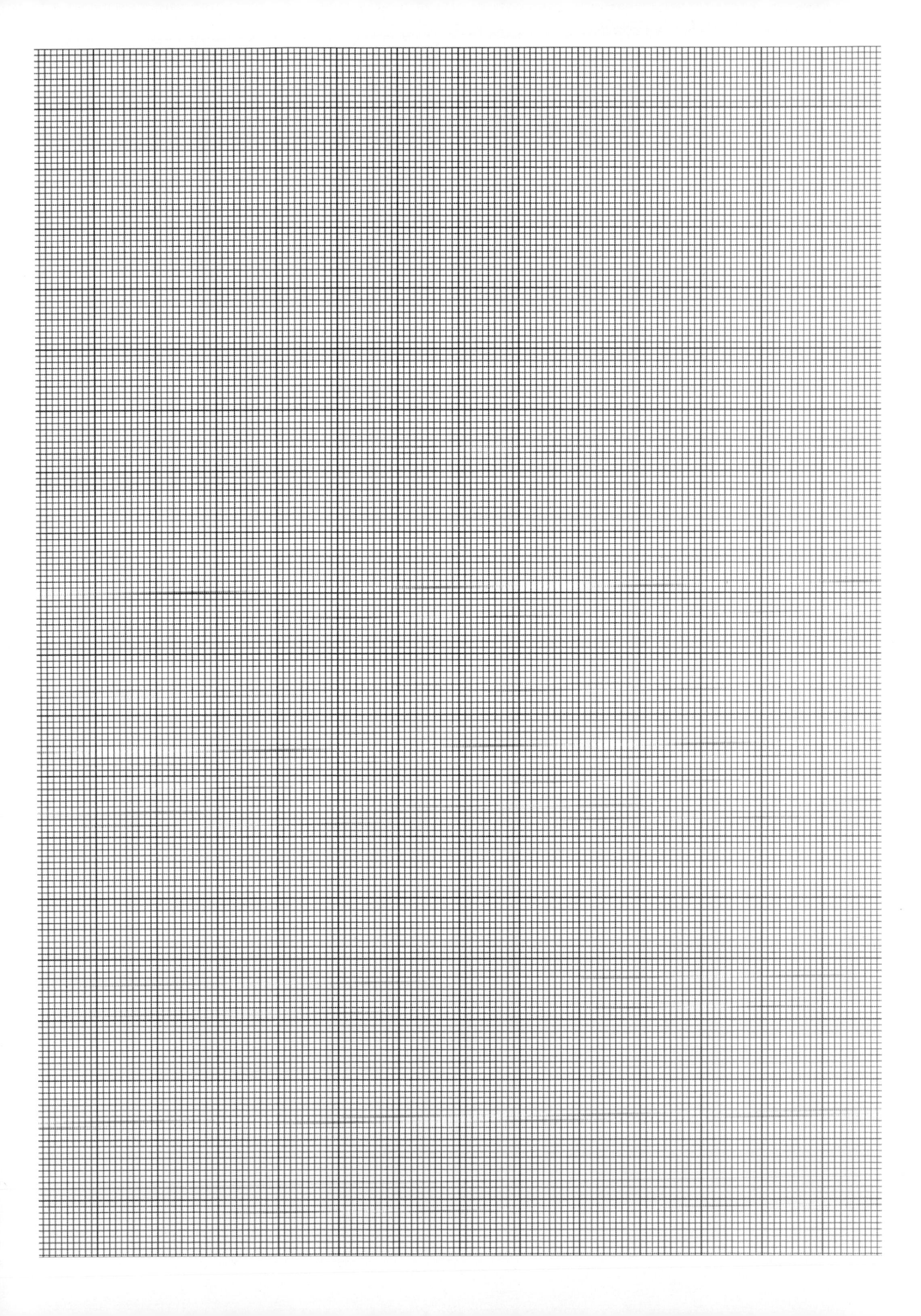

exercise

# FIFTEEN

## *Subsurface-Fluid Withdrawal and Ground Subsidence*

Ground subsidence as a consequence of subsurface-fluid withdrawal is a widespread geologic phenomenon. Three types of withdrawal can be responsible: (1) withdrawal of groundwater; (2) withdrawal of petroleum and natural gas; and (3) withdrawal of steam for geothermal power. Gentle subsidence bowls develop almost imperceptibly but can extend over very large areas, such as in the San Joaquin Valley of California, where a region 150 mi long and 30 mi wide has been affected. Several areas in the Valley have settled almost 30 ft since 1926 (figure 15.1). Other densely populated areas significantly affected by ground subsidence include New Orleans, Baton Rouge, Houston-Galveston (figure 15.2), Savannah, Mexico City, London, and Venice. Increasing, widespread exploitation of fluid natural resources, especially groundwater, is certain to multiply the incidences of ground subsidence. Predictable effects include bent and ruptured water and sewer pipes, cracked building foundations, surface faulting, and nearshore flooding.

Studies of many regions affected by ground subsidence during the past 60 years help us to evaluate numerous contributing causes:

1. Most important is the pumped decline of groundwater head. As the fluid level declines, pore-water pressure decreases, effective stress on detrital grains in the sand increases, and some compaction of the sand aquifer occurs. The compaction is accompanied by rearrangement of the grains and by increased closeness and plastic deformation of ductile grains such as clay, mica, and schist fragments. This process reduces both porosity and permeability, in addition to thinning the sand layer.
2. More important than sand-grain rearrangement as a cause of ground subsidence is the subsequent compaction of interbedded clay-rich confining beds, the shale aquitards. These beds, though not pumped themselves during water production, compact as pore water moves from the shale into the sandstone to equalize with the lower pore pressures induced in the adjacent sandy aquifer. In the San Joaquin Valley, for example, the specific subsidence (feet of subsidence per foot of rock) rises from 0.01 near the basin margins to 0.08 near the center, as coarse-grained alluvial fans give way to finer sediments farther from the mountain front. The widespread distribution of sand-shale sequences beneath alluvial plains throughout the world accounts for the great abundance of this type of subsidence, particularly because the shallow sand aquifers are easy to exploit and overpump.
3. An additional, important control over the amount of subsidence is the nature of the clay minerals in the shale confining beds. Montmorillonite, which contains a large amount of water, loses three times as much water under the new pore-pressure regime as kaolinite; illite is intermediate in water loss. The amount of subsidence in muds deposited within the past few thousand years correlates very well with the percentage of montmorillonite clay in the muds (table 15.1).

**Figure 15.1a**

Land subsidence in the San Joaquin Valley, California, 1926–1970. Most of the subsidence resulted from groundwater withdrawal, but some is due to petroleum removal as well.

*Source: R. L. Ireland, 1984, p. 4, U.S. Geological Survey.*

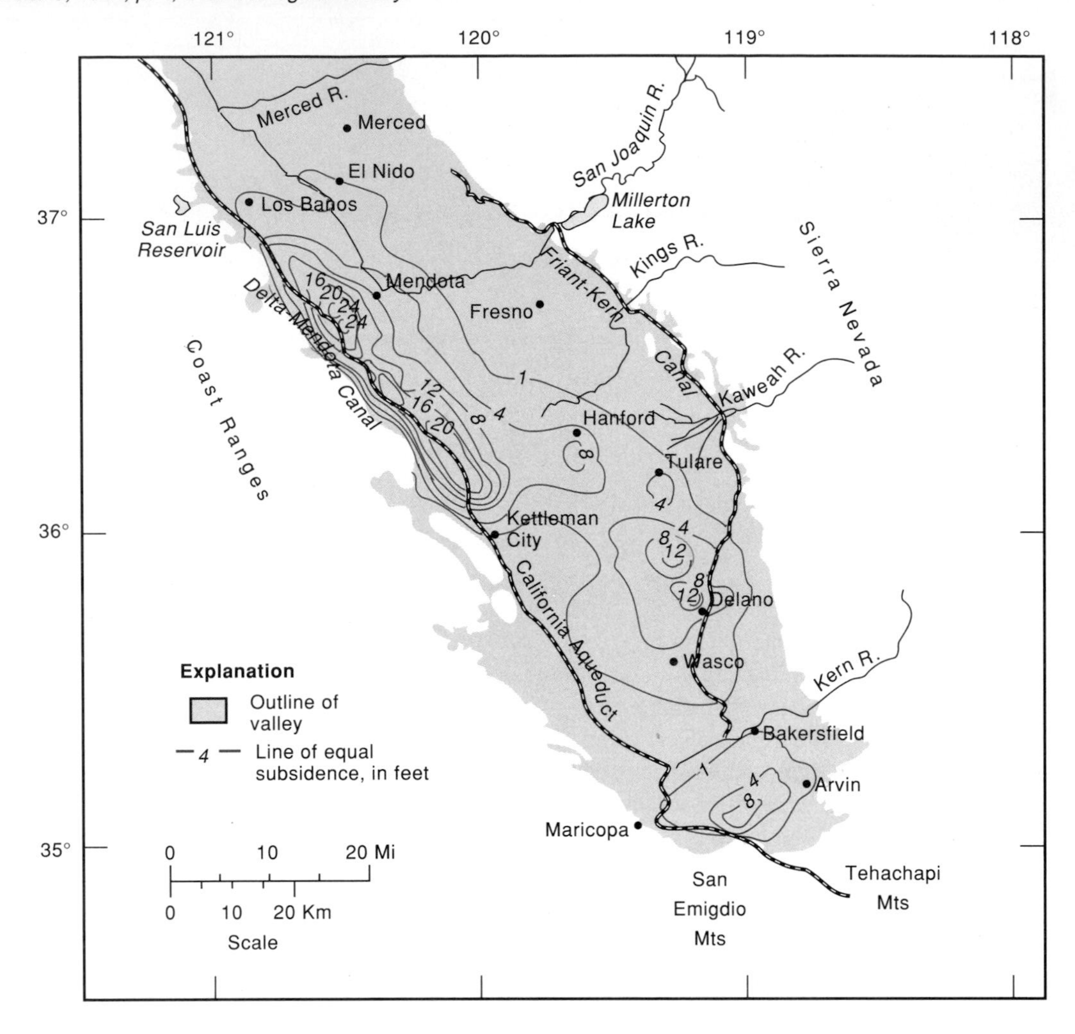


## TABLE 15.1

### Geological, Hydrological, and Subsidence Parameters at Four Sites

| Site | Clay Thickness (m) | Head Decline (m) | Subsidence (m) | Compressibility $\times 10^{-4}$ m/m/m | Montmorillonite Content (%) |
|---|---|---|---|---|---|
| Venice | 130 | 9 | 0.12 | 1.0 | 10 |
| Houston | 150 | 90 | 2.3 | 2.1 | 50 |
| Santa Clara | 145 | 49 | 5.3 | 7.4 | 70 |
| Mexico City | 50 | 55 | 9.0 | 32.0 | 80 |

Subsidence at Santa Clara is the predicted ultimate value, which has not yet been realized. All other figures are current values. The amount of ground subsidence is unrelated to either the thickness of the clay or the decline in head but correlates strongly with the percentage of montmorillonite in the clay.

*From A. C. Waltham,* Ground Subsidence. *Copyright © 1989 Blackie Academic & Professional (an imprint of Chapman & Hall), London and Glasgow. Reprinted by permission.*

**Figure 15.1b**

Magnitude of subsidence at a site 10 miles southwest of Mendota, San Joaquin Valley, California. Power pole shows position of land surface in 1925, 1955, and 1977. Land surface was lowered about 30 feet during that period.

Unfortunately, ground subsidence cannot be reversed, because compaction and consolidation of the clay-rich aquitards causes permanent bonding between adjacent clay flakes. Subsidence can be halted, however, by pumping water back into the sand aquifer. The difficulty in this

**Figure 15.2**

Surface subsidence in the Houston-Galveston area, Texas. The faults are a result of the subsidence. About 30 square miles of coastal land have been permanently flooded by the ground lowering, and the risk of flooding from the frequent hurricane winds now extends further inland. Subsidence also contributes to problems of shoreline erosion along the coastline.

*Source: U.S. Geological Survey* Annual Report, 1982.

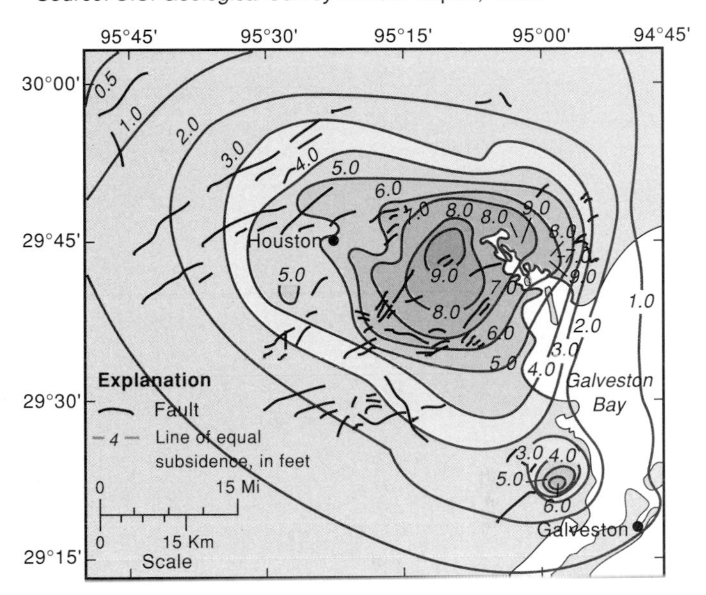


procedure is finding a sufficiently large water source. The most effective method is to decrease the rate at which water is removed from the aquifer and then reroute existing streams to increase the rate of replenishment of the aquifer.

Subsidence over petroleum and natural-gas fields is usually restricted by low compressibility of the lithified rocks at the great depths at which most hydrocarbon reservoirs occur. Clay-rich rocks buried many thousands of feet deep normally are as compacted as they can get, so decreasing fluid pressure by removing fluid from the oil-bearing sandstones has no effect on the shale beds, the beds responsible for most ground subsidence in areas of fluid withdrawal. Areas most vulnerable to significant subsidence are those where the petroleum and natural gas are in relatively young sediments at shallow depths, that is, where the geologic setting is similar to that in groundwater-withdrawal areas. Examples include a 14-foot subsidence over some of the oil fields around Lake Maracaibo, Venezuela, where the fluids are 1,000–4,000 ft deep, and about a 30-foot subsidence over the Wilmington oil field at Long Beach, California (figure 15.3), where oil

**Figure 15.3**

Aerial photograph of the coastal area around Long Beach, California. Withdrawal of petroleum from the Wilmington oil field resulted in ground subsidence of 30 feet.

is produced from several sands 2,500 to 6,000 feet deep. Because much of the Long Beach area is at or near sea level, dikes and sea walls were needed to protect structures from inundation. By 1960 subsidence had reached almost 30 feet, prompting remedial measures: water was injected into the peripheral areas of the underlying oil fields to raise the pressures in the fields and stop the pressure decline causing the subsidence. These measures were largely successful; little or no subsidence is now taking place in the Long Beach area.

Ground subsidence has also been recorded at some sites that exploit geothermal hot-water and steam resources. The Wairakei site in New Zealand has wells 2,000–4,000 feet deep that have created a subsidence bowl 1.2 mi$^2$ in area and 15 ft deep. Similarly, vertical displacements of about 5 inches have occurred at The Geysers geothermal field in Lake County, California, about 75 miles north of San Francisco. In The Geysers area the volume changes resulting from fluid withdrawal and subsidence also seem to have increased the frequency of microearthquakes. Comparison of seismic activity during preproduction (1962–1963) and peak production (1975–1977) times showed that regional seismicity (magnitude $\geq 2$) increased from 25 events per year in the former period to 47 per year in the latter. To reduce subsidence, steam

**Figure 15.4**

Measured differential subsidence in Jefferson Parish, Louisiana, related to peat thickness. Blank area in western part of map is undeveloped or currently being developed.

*From J. O. Snowdon,* Gulf Coast Association of Geological Societies Transactions, *27:179, 1977, Gulf Coast Association of Geological Societies. Used with permission.*

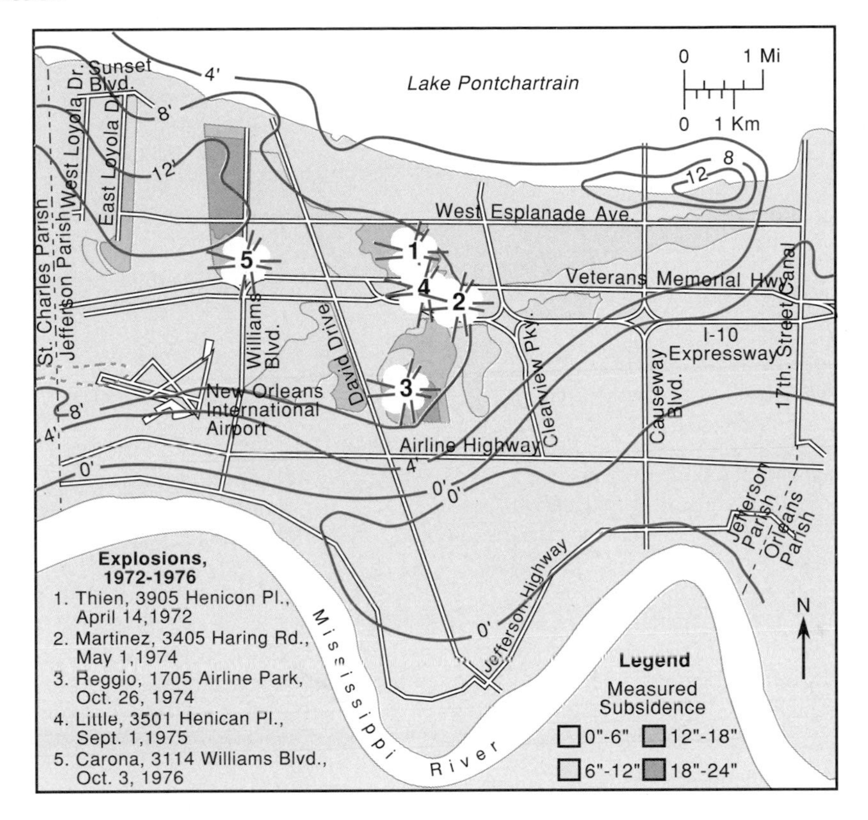


condensate was then reinjected into the producing formation, but this process might eventually increase seismic activity in this earthquake-prone region by inducing slippage along the many planes of weakness. Power generation at The Geysers apparently has hidden, unavoidable costs.

4. A large part of metropolitan New Orleans is built on marshland that contains up to 15 feet of marsh-grass peat, a poor substrate for construction. These organic-rich sediments are easily compressed, so significant ground subsidence has occurred and continues today (figure 15.4). The result has been flooding and widespread structural damage to sewer, water, and natural-gas lines and to streets, driveways, and sidewalks. Houses tilt, cement slabs crack, and building walls crack as well. Numerous natural-gas explosions have resulted from ruptured pipes.

## Problems

1. Most ground subsidence occurs in areas underlain by clastic sediments deposited within the past few tens of millions of years (the last half of the Tertiary Period). Explain why this is so.
2. Shown are data from the Inglewood oil field in Los Angeles County, California. The oil-producing horizon is of Pliocene and possibly Pleistocene age and consists of poorly consolidated marine silts and very fine-grained sands 1,000–2,000 ft deep. The table indicates the amount of liquid produced (oil plus associated salt water) and the volume of land subsidence for five time periods since oil production began in 1911.

| | **Liquid Production ($ft^3$)** | **Volume of Subsidence ($ft^3$)** |
|---|---|---|
| Nov. 1911–Oct. 1943 | 1,130,000,000 | 95,552,000 |
| Oct. 1943–Mar. 1950 | 567,600,000 | 39,900,000 |
| Mar. 1950–Aug. 1954 | 433,000,000 | 35,860,000 |
| Aug. 1954–Oct. 1958 | 376,600,000 | 27,020,000 |
| Oct. 1958–Aug. 1962 | 299,800,000 | 19,760,000 |

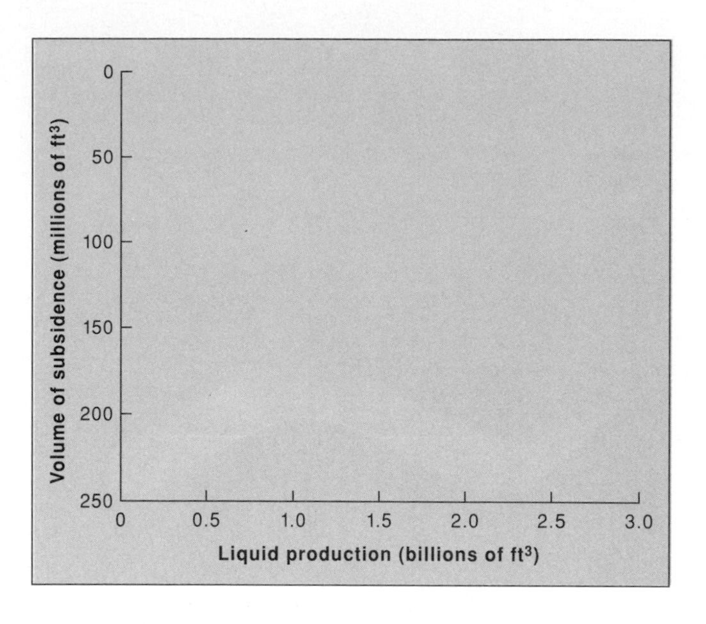


a. On the graph above, plot these two variables against each other and connect the five data points by lines. How do you interpret the shape of the line from 1943 to 1962?
b. Calculate the ratio between the volume of subsidence and liquid production for each of the five time periods. Are the five ratios similar? How might you interpret this result?

3. With reference to figure 15.1A, land subsidence in the San Joaquin Valley,
   a. What is the minimum average rate of subsidence per year in the Valley?
   b. Why do you think the areas of maximum subsidence are located near the coastline?
   c. Describe the likely effect of land subsidence in the San Joaquin Valley on the regime of rivers draining east-to-west from the Sierra Nevada or west-to-east from the Coast Ranges. How might erosion and sediment transport be affected?
   d. The San Joaquin Valley produces a large percentage of fruits and vegetables for the American population, a circumstance made possible by extensive irrigation. Restricting farmers' use of subsurface water would drive many of them into bankruptcy, with associated harmful effects to agriculture-based industries. What should the California state government do to balance the competing needs of the agricultural community and city-dwellers and factory-owners whose buildings are being destroyed by the subsidence?
4. Population density and oil/gas production are important factors in studies of ground subsidence. Examine a geologic map of the United States and identify as many locations as possible where significant ground subsidence can be expected.

## Further Reading/References

Bull, W. B., 1975, *Land Subsidence Due to Ground-water Withdrawal in the Los Banos–Kettleman City Area, California, Part 2. Subsidence and Compaction of Deposits.* U.S. Geological Survey Professional Paper 437-F, 90 pp.

Holzer, T. L. (ed.), 1984, Man-Induced Land Subsidence. *Reviews in Engineering Geology,* v. VI, Boulder, Colorado, Geological Society of America, 221 pp.

Ireland, R. L., Poland, J. F., and Riley, F. S., 1984, *Land Subsidence in the San Joaquin Valley, California, as of 1980.* U.S. Geological Survey Professional Paper 437-I, 93 pp.

Poland, J. F., and Davis, G. H., 1969, Land subsidence due to withdrawal of fluids: in *Reviews in Engineering Geology,* v. II, D. J. Varnes and G. Kiersch (eds.), Boulder, Colorado, Geological Society of America, pp. 187–269.

Poland, J. F., and Ireland, R. L., 1988, *Land Subsidence in the Santa Clara Valley, California, as of 1982.* U.S. Geological Survey Professional Paper 497-F, 61 pp.

Waltham, A. C., 1989, *Ground Subsidence.* New York, Chapman and Hall, 202 pp.

exercise

# SIXTEEN

# Soil Pollution

Soils are formed by the chemical and, to a much lesser extent, physical alteration of rocks and sediments exposed at the Earth's surface. Some soils are residual, remaining where they formed from underlying materials. Other soils have developed on alluvial debris deposited by ancient streams. But whether residual or alluvial in origin, all soils go through the same developmental stages and reflect the same variables: parent materials, climate, and time. Depending on the first two variables, one meter of soil might require between 100 and 100,000 years to form.

Soils are mixtures of sand, silt, and clay particles but there is no consensus regarding a nomenclature system for describing different proportions of these three particle sizes (figure 16.1). The most agriculturally productive soils contain subequal mixtures of them. Most important from the viewpoint of pollution and pollution control is the clay-size fraction, composed almost entirely of clay minerals, clay colloids, and organic matter. Much of the silt fraction of a soil can consist of clots or aggregates of these substances as well. The importance of these clay-size substances lies in their ability to adsorb both inorganic ions and polar and nonpolar organic compounds. This property serves to keep harmful substances from entering the groundwater supply below the soil, but also causes harm to plant life rooted in the soil, and to animals that eat the plants.

### *Structure of Clay Minerals*

Because clay minerals are critically important in soil pollution (and other environmental problems), environmental scientists must understand the behavior of these minerals and, therefore, their crystal structure (figure 16.2). Clay minerals are sheeted aluminosilicate minerals (minerals whose essential elements are aluminum, silicon, and oxygen). The sheets are of two types. One type is formed by silicon and oxygen ions arranged in a tetrahedral shape with the four oxygen ions at the four corners of the tetrahedron and the silicon ion in the center. These are $SiO_4^{-4}$ groups. Each tetrahedron shares its basal three oxygens with adjacent tetrahedra to form an essentially planar surface, the cleavage surface of the clay mineral. (Micas have the same structure.) The second type of sheet is composed of aluminum-hydroxyl groups arranged in octahedra. These octahedra link by sharing hydroxyls with adjacent octahedra. Kaolinite is a clay mineral composed of one Si-O sheet and one Al-OH sheet. Montmorillonite and illite are composed of three sheets, one Al-OH sheet situated between two Si-O sheets.

In addition to being very small, usually less than one micron in length, clay minerals have defective crystal structures. In illite most of the defects consist of the substitution of aluminum for about 25% of the silicon ions in the tetrahedral sheet. Because of the considerable size difference between silicon and aluminum ions (table 16.1), the clay structure cannot tolerate a substitution level greater than about 25%. In montmorillonite most of the defects result from substitution of $Fe^{+3}$, $Fe^{+2}$, and $Mg^{+2}$ for aluminum in the octahedral sheet. Kaolinite does not have significant substitution; chemical analysis reveals its metallic cations to be at least 99% silicon and aluminum.

Most of the substitutions in illite and montmorillonite create charge deficiencies ($Al^{+3}$ for $Si^{+4}$, $Mg^{+2}$ for $Al^{+3}$, $Fe^{+2}$ for $Al^{+3}$) that are balanced by cations adsorbed on the clay mineral surfaces. Illite prefers to adsorb potassium ions, while montmorillonite prefers sodium or calcium ions. Other cations might serve as well, but the cations used are those normally abundant in soil water and groundwater. When human activities add heavy metals or radioactive substances to the soil, these substances too can be adsorbed onto clay mineral surfaces to balance the internal charge deficiencies of the clay. Many naturally occurring organic compounds are also positively charged, and the bulk of organic pollutants (for example, PCBs) are polar molecules, each with a positively charged end that is attracted to clay-mineral surfaces. Negatively charged organic molecules such as those with carboxyl groups are not attracted in large numbers to clay surfaces because of the scarcity of negative-charge deficiencies on clay-minerals. (A few may exist at clay mineral edges.) Nonpolar organic molecules can be adsorbed onto clays by van der Waals forces (physical adsorption).

The number of adsorption sites on a clay mineral flake is limited by its number of charge deficiencies. Suppose two metallic cations are competing for the same site. Which will be adsorbed? Which is likely to remain in the soil water and perhaps enter the groundwater supply below? We can answer these questions by considering the charge and size of the ions. More highly charged cations displace less highly charged cations; within a group of cations with the same charge, the larger displace the smaller.

## *Soil Organic Matter*

Soils generally contain 1–6% organic matter by weight, an accumulation of partially disintegrated and decomposed plant and animal residues and other organic compounds synthesized by soil microbes as the decay occurs. In addition to these transitory materials, soil organic matter includes complex compounds, called *humus,* that are relatively resistant to decay. This material, usually brown or black, is colloidal in nature. For a given mass, humus can hold much more water and nutrient cations than can clay minerals, because of its amorphous (noncrystalline) nature. Amorphous compounds have greater ion-holding capacities than those of crystalline materials because adsorption is not

**Figure 16.1**

Three commonly used systems for naming mixtures of sand, silt, and clay.

*(a) Source: U.S. Department of Agriculture. (b) From R. L. Folk,* Journal of Geology, *62:349, 1954. Copyright © 1954 University of Chicago Press, Chicago, IL. Reprinted by permission. (c) From F. P. Shepard,* Journal of Sedimentary Petrology, *24:157, 1954. Copyright © 1954 SEPM (Society for Sedimentary Geology). Tulsa, OK. Reprinted by permission.*

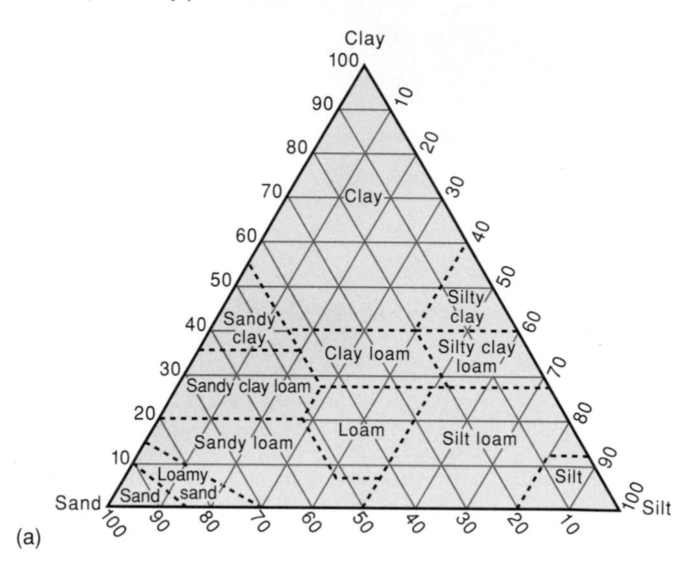

(a)

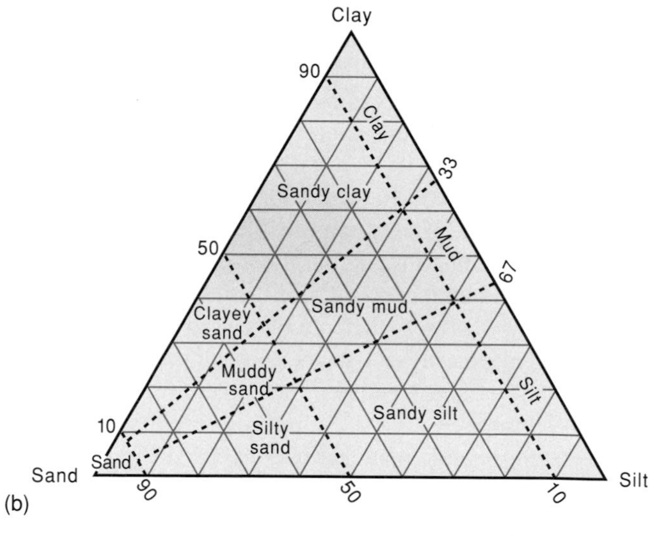

(b)

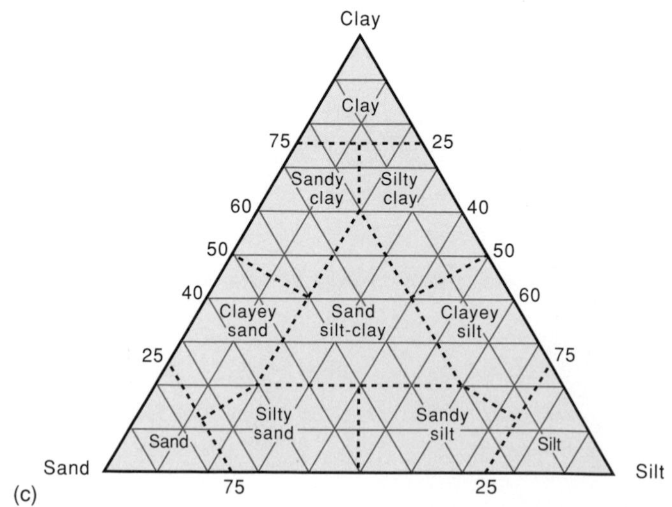

(c)

**Figure 16.2**

Diagrammatic sketches of (a) a kaolinite flake and (b) 1-1/3 illite or montmorillonite flakes, illustrating the layered crystal structure. Sorption of cations occurs on all planar (cleavage) surfaces.

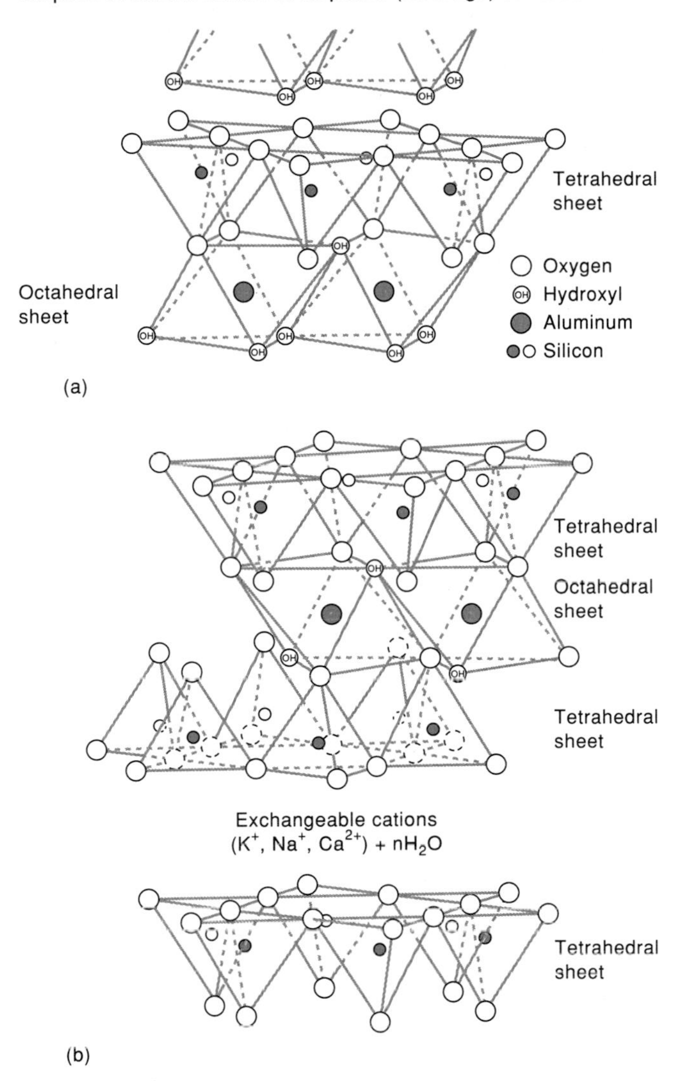


restricted to specific sites on each particle. Small amounts of humus thus greatly increase the soil's capacity to both promote plant growth and retain potentially harmful inorganic and organic pollutants.

## TABLE 16.1

### Ionic Radii of Naturally Occurring Elements

| Ion | Radius | Ion | Radius | Ion | Radius |
|---|---|---|---|---|---|
| $N^{+5}$ | 0.15 | $Li^{+1}$ | 0.68 | $Th^{+4}$ | 0.95 |
| *$C^{+4}$ | 0.16 | $Mo^{+4}$ | 0.68 | $Eu^{+3}$ | 0.97 |
| $B^{+3}$ | 0.20 | $W^{+4}$ | 0.68 | $Sm^{+3}$ | 0.97 |
| $S^{+6}$ | 0.29 | $As^{+3}$ | 0.69 | $Y^{+3}$ | 0.97 |
| $Be^{+2}$ | 0.34 | $Ti^{+3}$ | 0.69 | $Cu^{+1}$ | 0.98 |
| $P^{+5}$ | 0.35 | $Se^{+4}$ | 0.69 | *Na | 0.98 |
| $Cr^{+6}$ | 0.35 | $Mn^{+3}$ | 0.70 | $Pm^{+3}$ | 0.98 |
| $Se^{+6}$ | 0.35 | $V^{+2}$ | 0.72 | $Cd^{+2}$ | 0.99 |
| *$Si^{+4}$ | 0.39 | *$Mg^{+2}$ | 0.74 | $Nd^{+3}$ | 0.99 |
| $V^{+5}$ | 0.40 | $Ni^{+2}$ | 0.74 | $Pr^{+3}$ | 1.00 |
| $Ge^{+4}$ | 0.44 | $Bi^{+5}$ | 0.74 | $Sn^{+2}$ | 1.02 |
| $Mn^{+7}$ | 0.46 | $Pb^{+4}$ | 0.76 | $Ce^{+3}$ | 1.02 |
| $As^{+5}$ | 0.47 | $Co^{+2}$ | 0.78 | *$Ca^{+2}$ | 1.04 |
| $Mn^{+4}$ | 0.52 | $Ti^{+4}$ | 0.78 | $La^{+3}$ | 1.04 |
| $Re^{+6}$ | 0.52 | $Cu^{+2}$ | 0.80 | $U^{+3}$ | 1.04 |
| $Te^{+6}$ | 0.56 | *$Fe^{+2}$ | 0.80 | $Te^{+3}$ | 1.05 |
| *$Al^{+3}$ | 0.57 | $Lu^{+3}$ | 0.80 | $Pa^{+3}$ | 1.06 |
| $V^{+4}$ | 0.61 | $Yb^{+3}$ | 0.81 | $Th^{+3}$ | 1.08 |
| $Ga^{+3}$ | 0.62 | $Hf^{+4}$ | 0.82 | $Ac^{+3}$ | 1.11 |
| $Ru^{+4}$ | 0.62 | $Zr^{+4}$ | 0.82 | $Hg^{+2}$ | 1.12 |
| $Sb^{+5}$ | 0.62 | $Cr^{+2}$ | 0.83 | $Ag^{+1}$ | 1.13 |
| $Co^{+3}$ | 0.64 | $Zn^{+2}$ | 0.83 | $Sr^{+2}$ | 1.20 |
| $Cr^{+3}$ | 0.64 | $Sc^{+3}$ | 0.83 | $Bi^{+3}$ | 1.20 |
| $Pt^{+3}$ | 0.64 | $Er^{+3}$ | 0.85 | $Pb^{+2}$ | 1.26 |
| $Pd^{+4}$ | 0.64 | $Tm^{+3}$ | 0.85 | *$K^{+1}$ | 1.33 |
| $Ti^{+4}$ | 0.64 | $Ho^{+3}$ | 0.86 | $F^{-1}$ | 1.33 |
| $Ge^{+2}$ | 0.65 | $Dy^{+3}$ | 0.88 | *$O^{-2}$ | 1.36 |
| $Ir^{+4}$ | 0.65 | $Ce^{+4}$ | 0.88 | $Au^{+1}$ | 1.37 |
| $Os^{+4}$ | 0.65 | $Tb^{+3}$ | 0.89 | $Ba^{+2}$ | 1.38 |
| $Rh^{+4}$ | 0.65 | $U^{+4}$ | 0.89 | $Ra^{+2}$ | 1.44 |
| $Mo^{+6}$ | 0.65 | $Te^{+3}$ | 0.89 | $Rb^{+1}$ | 1.49 |
| $W^{+6}$ | 0.65 | $Sb^{+3}$ | 0.90 | $Tl^{+1}$ | 1.49 |
| $Nb^{+5}$ | 0.66 | $La^{+4}$ | 0.90 | $Cs^{+1}$ | 1.65 |
| $Ta^{+5}$ | 0.66 | $Mn^{+2}$ | 0.91 | $Cl^{-1}$ | 1.81 |
| *$Fe^{+3}$ | 0.67 | $Pa^{+4}$ | 0.91 | $S^{-2}$ | 1.82 |
| $V^{+3}$ | 0.67 | $In^{+3}$ | 0.92 | $Br^{-1}$ | 1.96 |
| $Nb^{+4}$ | 0.67 | $Gd^{+3}$ | 0.94 | $I^{-1}$ | 2.20 |
| $Sn^{+4}$ | 0.67 | | | | |

Excluding $H^{+1}$, $He^{+2}$, Noble Gases, and Elements with no stable nuclide: Tc (43), Po (84), At (85), Fr (87)

Radii in angstroms. One angstrom = $10^{-10}$ m.

*Carbon and the eight other most abundant elements in the Earth's crust.

### *Cation-Exchange Capacity*

Ions or polar molecules held on clay-mineral and humic substances can be replaced by other ions or polar molecules; the degree of replacement possible is termed the cation-exchange capacity (CEC) of the clay. The CEC depends on the number of charge deficiencies within the clay so, as montmorillonite is the finest-grained clay and has the most charge deficiencies, it has the highest CEC. Kaolinite, because it has few or no substitutions for the aluminum and silicon in its crystal structure, has few charge deficiencies and the lowest CEC. Cation-exchange capacity is expressed as the number of moles (or equivalents) of a monovalent cation that can be replaced. Thus, if a soil has a

CEC of 10 mol/kg, 1 kg of the soil can adsorb 10 mol of $H^+$, for example, and exchange it for 10 mol of another monovalent cation such as $Na^+$ or $K^+$, or for 5 mol of a divalent cation such as $Ca^{+2}$, or for 3 1/3 mole of $Al^{+3}$. Representative CECs are kaolinite, 5 mmol/100 g; illite, 25 mmol/100 g; montmorillonite, 90 mmol/100 g; and humus, 200 mmol/100 g. Suppose a soil contains 10% montmorillonite as its only particles with significant cation-exchange capacity. How much lead can it adsorb?

$$10\% \text{ montmorillonite} \times 90 \text{ meq/100 g} \times 2 \text{ (the charge of } Pb^{+2}) = 18 \text{ meq/100 g soil}$$

$$18 \text{ mmol Pb/100 g } (207 \text{ g/mol} \times 1/1000 \text{ mol/mmol}) = 3.73\% \text{ Pb} = 37{,}300 \text{ ppm Pb}$$

When interpreting the results of these calculations, we must keep several considerations in mind.

1. A wide range of CEC values is possible for each type of clay mineral and humus. Kaolinite varies from 3 to 20 mmol/100 g, illite from 10 to 40, montmorillonite from 80 to 120, and humus from 135 to 300. Without detailed mineralogical and chemical analyses of the soil clay, CEC calculations can be only approximations, in error by perhaps a factor of two.
2. The calculation is meaningful only if the potentially harmful element occurs in an available form in the soil. If lead, for example, is present in an insoluble or inaccessible form, such as inclusions of lead sulfide (galena) in quartz grains, the calculation is pointless.
3. Mineral solubilities can change by orders of magnitude in response to changes in pH or oxidation-reduction potential, greatly changing the availability of an element.

Some elements that are essential for plant life in small amounts are harmful when present in much greater amounts, but the tolerances of many species for different elements is not known. The same may be said of animal and human tolerances. Table 16.2 gives common concentrations of some elements in soils and plants. Table 16.3 gives the ranges of these elements and many others in materials that humans often add to soils.

TABLE 16.2

Range of Concentration in Soils and Plants of Some Elements that Sometimes Occur as Environmental Contaminants

| | Common Range in Concentration (ppm) | |
|---|---|---|
| **Element** | **Soils** | **Plants** |
| Arsenic | 0.1–40 | 0.1–5 |
| Boron | 2–100 | 30–75 |
| Cadmium | 0.1–7 | 0.2–0.8 |
| Copper | 2–100 | 4–15 |
| Lead | 2–200 | 0.1–10 |
| Manganese | 100–4,000 | 15–100 |
| Nickel | 10–1,000 | 1 |
| Zinc | 10–300 | 15–200 |

*Source: Data from N. C. Brady,* The Nature and Properties of Soils, *8th edition, 1974, Macmillan Publishing Company, New York, NY.*

## TABLE 16.3

Typical Ranges of Heavy Metal Concentrations in Sewage Sludges, Fertilizers, Farmyard Manure, Lime, and Composts (mg/kg)

| | Sewage Sludge | Phosphate Fertilizers | Nitrate Fertilizers | Farmyard Manure | Lime | Composted Refuse |
|---|---|---|---|---|---|---|
| Ag | <960 | — | — | — | — | — |
| As | 3–30 | 2–1,200 | 2.2–120 | 3–25 | 0.1–25 | 2–52 |
| B | 15–1,000 | 5–115 | — | 0.3–0.6 | 10 | — |
| Cd | <1–3,410 | 0.1–170 | 0.05–8.5 | 0.1–0.8 | 0.04–0.1 | 0.01–100 |
| Co | 1–260 | 1–12 | 5.4–12 | 0.3–24 | 0.4–3 | — |
| Cr | 8–40,600 | 66–245 | 3.2–19 | 1.1–55 | 10–15 | 1.8–410 |
| Cu | 50–8,000 | 1–300 | — | 2–172 | 2–125 | 13–3,580 |
| Hg | 0.1–55 | 0.01–1.2 | 0.3–2.9 | 0.01–0.36 | 0.05 | 0.09–21 |
| Mn | 60–3,900 | 40–2,000 | — | 30–969 | 40–1,200 | — |
| Mo | 1–40 | 0.1–60 | 1–7 | 0.05–3 | 0.1–15 | — |
| Ni | 6–5,300 | 7–38 | 7–34 | 2.1–30 | 10–20 | 0.9–279 |
| Pb | 29–3,600 | 7–225 | 2–27 | 1.1–27 | 20–1,250 | 1.3–2,240 |
| Sb | 3–44 | <100 | — | — | — | — |
| Se | 1–10 | 0.5–25 | — | 2.4 | 0.08–0.1 | — |
| U | — | 30–300 | — | — | — | — |
| V | 20–400 | 2–1,600 | — | — | 20 | — |
| Zn | 91–49,000 | 50–1,450 | 1 42 | 15–566 | 10–450 | 82–5,894 |

## Problems

1. A soil scientist has determined that for a wide variety of soil types, from a wide range of climates, cation-exchange capacity (CEC) is related simply to the quantities of clay and organic matter present in the soil:

   CEC = 0.57 (% clay) + 4.55 (% organic carbon).

   a. Calculate CEC values for the following combinations of clay and organic matter.

   | Clay % | Organic Matter % |
   |---|---|
   | 10 | 0.5 |
   | 10 | 5.0 |
   | 30 | 0.5 |
   | 30 | 5.0 |

   b. What is the quantitative difference in importance between clay and organic matter when the two are present in equal amounts?
   c. Approximately what is the quantitative difference in CEC if a soil clay is composed entirely of montmorillonite rather than kaolinite?
   d. Almost all soils have a much higher percentage of clay than of organic matter. Does this fact necessarily reflect the relative contributions of these two constituents to the cation-exchange capacity of a soil? Explain.
2. A soil has a cation-exchange capacity of 15 meq/100 g. Is there enough CEC to absorb waste cadmium at the rate of 5 tons/acre [1/2 (lb/acre) = ppm]? The atomic weight of divalent cadmium is 112.
3. Which type of clay mineral is most likely to be abundant in acidic soils? Explain.
4. Why do you think trivalent ions displace divalent ions and divalent ions displace monovalent ions? (Hint: consider their positions in the periodic table.)
5. Which of the heavy metals listed in table 16.3 would attach to clay mineral surfaces most strongly? Can you think of any ways to remove them from the clays?
6. Lime (powdered calcium carbonate) can contain very high amounts of manganese and lead but only very low amounts of cadmium, mercury, or boron. For each element, explain why this is so.
7. In what way does the amount of humus in a soil affect the availability of iron to plant life?
8. Construction workers have added large amounts of slag (glass from a smelting operation) containing 2,000 ppm lead to a residential soil to level the uneven ground surface. What factors must be considered to evaluate the likelihood that the potentially harmful lead ions will be freed from the slag and adsorbed onto soil material?
9. Suppose the slag contained uranium rather than lead. In what ways would this change your approach to answering the previous question?
10. In table 16.2 it is clear that plants generally contain lesser amounts of the eight elements listed than does the soil in which the plants grow. Based on this information, what might you infer about the chemistry of plant growth? Design an experiment to determine the tolerance of a plant to arsenic, listing the factors you must consider.

## Further Reading/References

Aiken, G. R., MacKnight, D. M., Wershaw, R. L., and MacCarthy, P. (eds.), 1985, *Humic Substances in Soil, Sediment, and Water*. New York, John Wiley & Sons, 692 pp.

Alloway, B. J. (ed.), 1990, *Heavy Metals in Soils*. New York, John Wiley & Sons, 339 pp.

Brady, N. C., 1990, *The Nature and Properties of Soils,* 10th ed. New York, Macmillan, 621 pp.

Paton, T. R., 1978, *The Formation of Soil Material*. Boston, George Allen & Unwin, 143 pp.

Sawhney, B. L., and Brown, K. (eds.), 1989, *Reactions and Movement of Organic Chemicals in Soils*. Madison, Wisconsin, Soil Science Society of America Special Publication No. 22, 474 pp.

# exercise SEVENTEEN

## Construction Stone

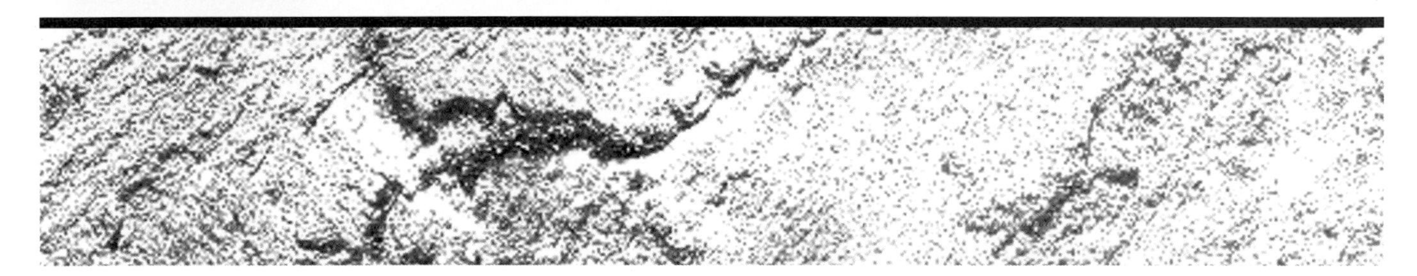

If asked to name the most important materials removed from the Earth for human use, most people probably would name petroleum, natural gas, and coal. Few would cite cut and crushed stone, sand and gravel, and clay. Yet these materials are essential for constructing commercial buildings, homes, highways, airport runways, and all other artificial structures. Sand and gravel is the only mineral commodity produced in all 50 states; crushed stone is mined in 48 states. If the annual use of these materials is apportioned among the population of the United States, each person "uses" 9,140 lbs of stone, 8,580 lbs of sand and gravel, 800 lbs of cement, and 490 lbs of clay. The clay is used to manufacture bricks and as an additive in cement, which is made mostly of finely ground limestone. These four commodities total more than 90% of the nonmetallic natural materials used each year.

These nonmetallics are composed mostly of a few rock and mineral types: limestone, dolostone, quartzite, sandstone, coarse-grained igneous rocks such as granites, and quartz sand. The carbonate rocks are used largely for cement and as aggregate in concrete, although small amounts are used as facing stone on buildings. For use in aggregate, limestone and dolostone are mined (quarried) from their rock layers, and then crushed to appropriate sizes and sieved (screened). Quartzite and quartz sand are mined as stream-transported, unlithified gravel and sand and rarely require crushing, although they sometimes require screening.

What are the general requirements for establishing a quarry site for limestone or sand and gravel?

1. The raw material must be near enough to the ground surface and abundant enough to justify the cost of opening up shop at the location. Overhead costs include those of acquiring the property, mining the material, and transporting it to where it will be used.
2. The quarry site should be adjacent to major rail and road arteries and perhaps near a navigable river as well. At present, moving crushed rock or sand by truck costs about $0.10/mi/ton. Hence, the quarry must be no more than 50 miles from a city; the closer, the better.
3. The limestone rock or quartz-rich sediment must be pure enough for its intended use. For limestone this might mean at least 90% $CaCO_3$ with limited amounts of magnesium, iron, and silica, its three common contaminants (present as dolomite, hematite, and chert). Small amounts of clay are permissible, and perhaps even desirable if the limestone is to be used to make cement.

For quartz sand and gravel, the sediment should be free of clay minerals, mica, and organic matter. Impurities such as opal, chert, and most types of volcanic rocks are particularly undesirable, as they react chemically in portland cement (concrete) and cause it to expand, crack, and deteriorate. An ideal commercial sand and gravel deposit contains 60% gravel and 40% sand, providing ample coarse material to crush for road base or bituminous aggregate, and sand in the correct sizes and proportions for use in concrete. Substandard deposits can be upgraded by screening, washing, and combining sizes from different deposits.

Specifications controlling sand and gravel quality are highly variable, depending on the availability of materials and the purpose for which the aggregate is to be used. Specifications are normally concerned with (1) the reaction of the aggregates to alternate cycles of freezing and thawing or wetting and drying, both with and without salts present; (2) chemical reactivity; (3) resistance to abrasion and impact; (4) gradations in character of the aggregate; and (5) miscellaneous harmful contaminants. Water absorption, specific gravity, color, strength in fabricated concrete, and other characteristics can be important in particular regions or for particular purposes.

## *Origin of Commercial Deposits*

Commercially valuable limestone deposits are well lithified. Limestones form 10–15% of all sedimentary rocks and are, therefore, not difficult to locate. Not all limestones, however, are pure enough or close enough to cities to be commercially useful.

At a typical limestone quarry, the rock is blasted down and fed to crushers (figure 17.1); the crushed stone is then separated into various sizes by screening. Limestone's structure of interlocking calcite grains makes the fragments strong and gives them a high resistance to freezing and thawing. The stone does not produce enough quartz dust to abrade quarry machinery or scar workers' lungs.

Most commercial sand and gravel deposits are of Quaternary age because older deposits tend to be at least partially lithified. High-volume, high-velocity streams are the major environment of transportation and deposition of commercial gravel deposits. Wind moves sand but cannot transport gravel. Waves can produce and sort sediments of gravel size, but modern deposits of this type are limited to narrow strips along shorelines. Only stream waters move and concentrate large volumes of sand-gravel sediment. In addition, fast-moving streams act as grinding and washing mills, grinding up and removing soft, structurally weak, and commercially undesirable fragments and concentrating hard, sound ones. The waters also produce some amount of desirable rounding on grains coarser than about 1–2 mm.

Within a fluvial complex, commercial sand and gravel deposits are most likely either in braided streams or in the faster-moving, channel-center waters of nonbraided streams. Most such commercial deposits occur in one of two types of settings. In glaciated areas of the central and eastern United States, suitable deposits are abundant just south of the farthest advance of Pleistocene ice masses (figure 17.2). Here, the sand and gravel consists of debris that was carried by the glacier, dropped as the ice melted, and then transported outward by meltwater, the transportation serving to remove commercially undesirable silt and clay dumped by the glacier. In unglaciated areas the best deposits are located adjacent to highlands, such as the Coast Ranges and Sierra Nevada in California. In general, the closer to the highlands the better, because gravels are seldom transported great distances away from them. Because of both its mountain ranges and its rapid growth, California produces more sand and gravel than does any other state.

**Figure 17.1a**

A blast at a stone quarry.

**Figure 17.1b**

Broken limestone after a blast, ready for trucking to primary crushers.

## *Dimension Stone*

A few limestone deposits consist of thick, structureless (massive) beds without fractures or other partings (figure 17.3). Such deposits yield blocks that can be sawed, smoothed, turned, or carved to make buildings or monuments. Blocks are carefully cut by channeling machines and loosened from the quarry floor. From there they are taken to a mill, where they are fashioned according to specifications from an architect or designer. Such dimension stone is used as construction blocks, as facing on building exteriors, and as floors, panels, and windowsills. Many other types of igneous, sedimentary, and metamorphic rocks are also quarried for use as dimension stone. As with limestone, ease of quarrying, location near a large city and

## Figure 17.2a

Map showing glacially deposited sand and gravel availability in the area of the Independence fan, a glacial outwash feature 20 miles downstream from Lafayette, Indiana.

*Source:* Proceedings of the 18th Forum on the Geology of Industrial Minerals, 1982.

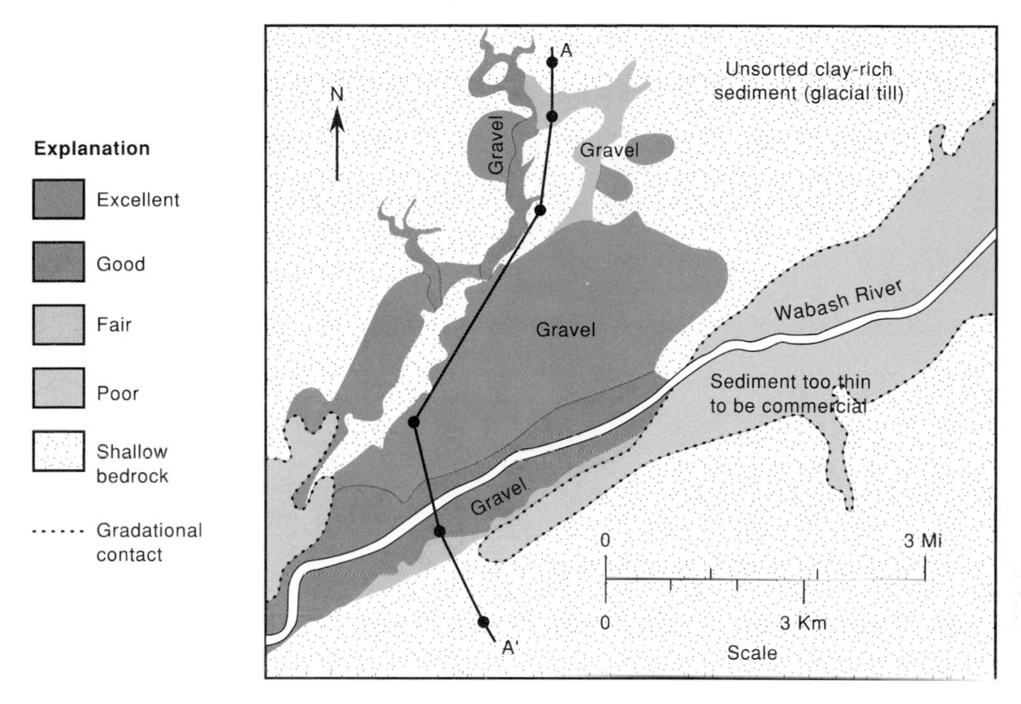


## Figure 17.2b

Cross-section along line A-A′ showing the unconsolidated deposits associated with the fan.

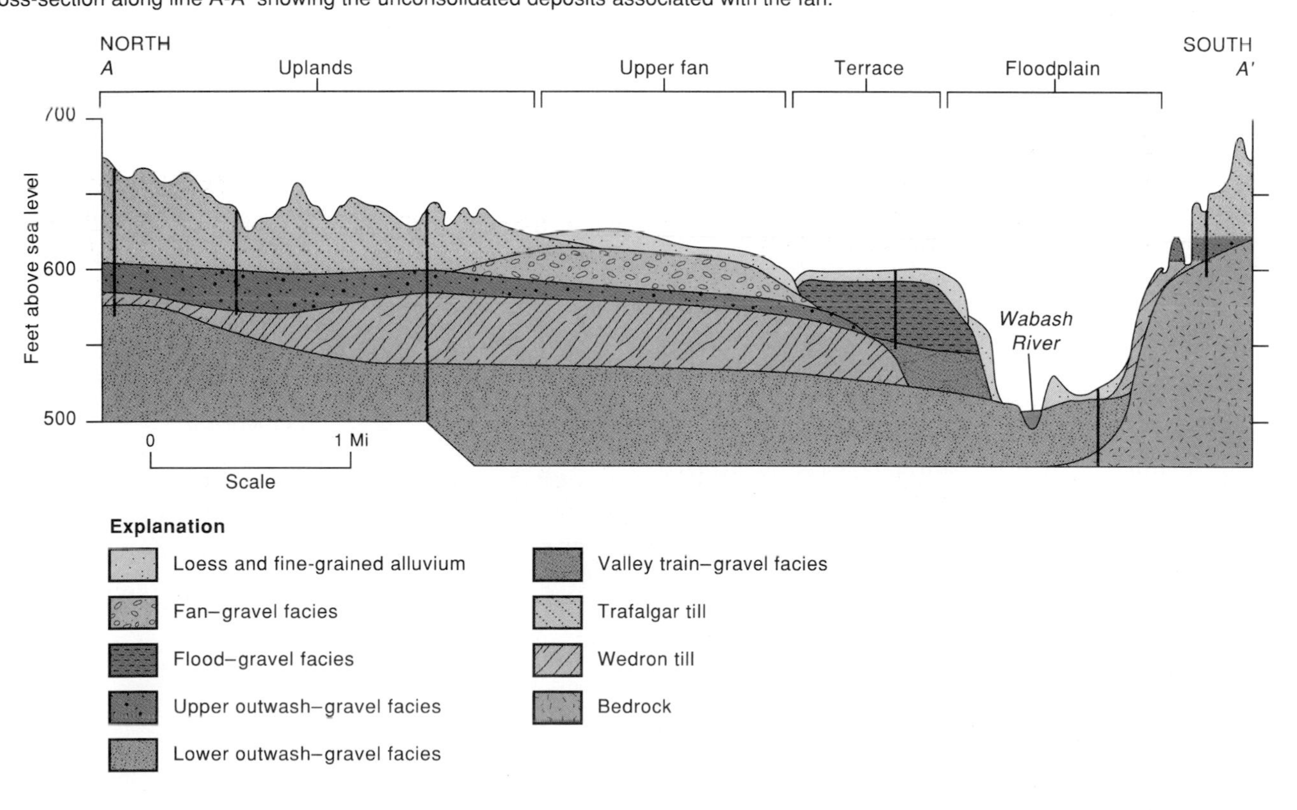

**Figure 17.3**

Quarrying Indiana limestone. The straight cuts are made by channeling machines (beyond the derrick). Long blocks, freed at the bottom, are "turned down" and split into smaller blocks for removal to the finishing plant.

transportation route, and physical attractiveness determine commercial value. The more attractive and desirable the stone, the less critical is location. Some types of dimension stone are transported many thousands of miles because of their desirability; well-known examples are Italian marble and travertine, an unusual variety of limestone.

## *Stone Quarries and Environmental Degradation*

Residential communities often view a stone quarry as an objectionable nuisance. Quarrying can be noisy because of dynamite blasting, can create significant truck traffic in areas that otherwise see only a few cars each day, and can produce unsightly blemishes on the landscape. In a typical case these problems occur because the quarry was started many decades earlier, when the site was perhaps 10 mi from the city limits. Since then the city has expanded, perhaps surrounding the quarry with expensive suburban homes. Former cornfields are now the site of a golf course and an artificial lake, both of which receive limestone dust from the blasting and drilling.

The quarry owners are, of course, aware that they are at the bottom end of the local popularity scale. But what can they do to please the residents of the peaceful suburban community? The owners believe they have been good corporate citizens for many years, supplying an essential commodity at a fair price. Perhaps half the concrete in the county was made using aggregate from the quarry. They have landscaped and fenced the quarry to keep out youngsters, schedule blasting as infrequently as possible, and distribute the blasting schedule in advance. Their trucks travel only on designated streets, at hours selected to minimize the impact on residential traffic.

Nevertheless, the company's days at the site are numbered. Although the company owns enough land to operate for many more years, a strongly backed movement is afoot to rezone the property from "industrial" to "commercial" or "residential." The county zoning commission is likely to approve this rezoning request, giving the quarry operators perhaps 12 months to cease operations. The community, which has been able to grow at will because of cheap local construction materials, is now in conflict with the industry that made that very growth possible. Possibly the quarry owners will be able to open a new site still beyond city limits (at least, for now), but they will have to pass on their higher transportation costs to consumers. This is not an uncommon story; many such companies are being zoned out of business in high-growth areas such as southern California and the New York City region.

## *Durability of Construction Stone*

In at least one characteristic, rocks are like people: they vary in strength. Some people can support a weight of 100 lbs on their shoulders while others can support only 50 lbs; similarly, blocks of rock have different bearing strengths. In construction work, two types of physical strength are important: compressive strength and tensile strength. Compressive strength is the weight required to crush the rock; tensile strength is the weight required to break a block of rock supported on the ends but not in the center. Typical strengths for common igneous, sedimentary, and metamorphic rock types are shown below.

| Rock Type | Compressive Strength ($lb/in^2$) | Tensile Strength ($lb/in^2$) | Ratio Between Maximum Compressive and Maximum Tensile Strength |
|---|---|---|---|
| Gabbro, diorite | 14000–49000 | 2100–4200 | 11.7 |
| Quartzite | 21000–42000 | 1400–4200 | 10.0 |
| Basalt | 21000–42000 | 1400–4200 | 10.0 |
| Granite | 14000–35000 | 980–3500 | 10.0 |
| Marble | 14000–35000 | 980–2800 | 12.5 |
| Limestone | 4200–35000 | 700–3500 | 10.0 |
| Slate | 14000–28000 | 980–2800 | 10.0 |
| Gneiss | 7000–28000 | 700–2800 | 10.0 |
| Sandstone | 2800–23800 | 560–3500 | 6.8 |

**Figure 17.4**

Chemical weathering of Cleopatra's Needle, a granite obelisk sculpted in 1600 BC. (a) Before it was removed from Egypt. (b) After a span of 75 years in New York City's Central Park. After surviving intact for about 35 centuries in Egypt, the windward side has been almost completely defaced in less than a century.

(a)

(b)

Compressive strengths of rocks are about ten times greater than their tensile strengths, so stone structures must be supported from below to prevent rupture. The arch, so prominent in ancient Roman construction, was developed largely because of this fact. Without buttressing on the sides, the top of the structure could not support itself.

Once a structure is completed, rock surfaces exposed to the atmosphere begin to change, usually in unsightly ways. These changes are caused by mechanical disintegration and chemical alteration of the rock. Numerous destructive processes have been identified. For example, new mineral crystals can grow within the pores on the outer surfaces of the rock, causing tiny rock fragments to spall off. In a similar process, water in cold climates enters rock pores and freezes as the temperature drops below 32°F. The water expands about 9% when it turns to ice, exerting more than enough pressure to fracture the rock. Biologic agents also are potent destructive forces. Ever-present bacteria, for example, produce acidic chemical secretions that deface once-attractive stone surfaces.

Rock is attacked chemically by both natural atmospheric agents and manufactured pollutants (figure 17.4). The most abundant minerals in building stones include quartz, the various types of feldspars, biotite and muscovite micas, and calcite. Each mineral reacts differently to chemical attack, forming a specific group of new substances and reacting at a rate different from those of other minerals. The rate of reaction under atmospheric conditions increases from quartz to sodium feldspar, to potassium feldspar, to muscovite mica, to calcic plagioclase, to biotite mica, to calcite. Both observations of outcrops and laboratory experiments indicate the following chemical reactions to be characteristic:

| $SiO_2$ | + | $2\,H_2O$ | $\rightleftharpoons$ | $H_4SiO_4$ |
|---|---|---|---|---|
| quartz | | water | | dissolved silica |

| $2\,NaAlSi_3O_8$ | + | $2\,H^+$ | + | $9\,H_2O$ | $\rightleftharpoons$ | $Al_2Si_2O_5(OH)_4$ | + | $4\,H_4SiO_4$ | + | $2\,K^+$ |
|---|---|---|---|---|---|---|---|---|---|---|
| sodium plagioclase feldspar | | hydrogen ion | | water | | kaolinite clay | | dissolved silica | | potassium ion |

| $2\,KAlSi_3O_8$ | + | $2\,H^+$ | + | $9\,H_2O$ | $\rightleftharpoons$ | $Al_2Si_2O_5(OH)_4$ | + | $4\,H_4SiO_4$ | + | $2\,K^+$ |
|---|---|---|---|---|---|---|---|---|---|---|
| orthoclase feldspar | | hydrogen ion | | water | | kaolinite clay | | dissolved silica | | potassium ion |

| $2\,KAl_3Si_3O_{10}(OH)_2$ | + | $2\,H^+$ | + | $3\,H_2O$ | $\rightleftharpoons$ | $3\,Al_2Si_2O_5(OH)_4$ | + | $2\,K^+$ |
|---|---|---|---|---|---|---|---|---|
| muscovite mica | | hydrogen ion | | water | | kaolinite clay | | potassium ion |

| $2\,CaAl_2Si_2O_8$ | + | $10\,H^+$ | + | $H_2O$ | $\rightleftharpoons$ | $Al_2Si_2O_5(OH)_4$ | + | $2\,H_4SiO_4$ | + | $2\,Ca^{+2}$ |
|---|---|---|---|---|---|---|---|---|---|---|
| calcium plagioclase | | hydrogen ion | | water | | kaolinite clay | | dissolved silica | | calcium ion |

| $2\,KFeMg_2AlSi_3O_{10}(OH)_2$ | + | $8\,H^+$ | + | $4\,H_2O$ | $\rightleftharpoons$ | $Al_2Si_2O_5(OH)_4$ | + | $4\,H_4SiO_4$ | + | $Fe_2O_3$ | + | $2\,K^+$ | + | $4\,Mg^{+2}$ |
|---|---|---|---|---|---|---|---|---|---|---|---|---|---|---|
| biotite mica | | hydrogen ion | | water | | kaolinite clay | | dissolved silica | | hematite | | potassium ion | | magnesium ion |

| $CaCO_3$ | + | $H^+$ | $\rightleftharpoons$ | $Ca^{+2}$ | + | $HCO_3^-$ |
|---|---|---|---|---|---|---|
| calcite | | hydrogen ion | | calcium ion | | bicarbonate ion |

Hydrogen ions are found dissolved in normal rainwater because some atmospheric carbon dioxide dissolves in rain as it falls through the atmosphere:

| $H_2O$ | + | $CO_2$ | $\rightleftharpoons$ | $H_2CO_3$ | $\rightleftharpoons$ | $H^+$ | + | $HCO_3^-$ | $\rightleftharpoons$ | $2\,H^+$ | + | $CO_3^{-2}$ |
|---|---|---|---|---|---|---|---|---|---|---|---|---|
| water | | carbon dioxide | | carbonic acid | | hydrogen ion | | bicarbonate ion | | hydrogen ion | | carbonate ion |

Hydrogen ions are the reason rainwater is acidic. Uncontaminated rainwater has a pH of 5.6, where neutral pH is 7.0 (pH = – log $H^+$ concentration). Natural rainwater thus contains 25 times more of the chemically aggressive hydrogen ions than does a neutral solution.

The precipitation called acid rain can contain one thousand times or more hydrogen ions than does natural precipitation; its pH values can be as low as 1 or 2, much more acidic than soft drinks or beer, whose acidity produces a tingling sensation in your mouth. The increased acidity of acid rain results primarily from burning fossil fuels in factories and automobiles; the rain in most industrialized parts of the world is now much more acidic than in undeveloped areas. In many areas of the world acid rain has had disastrous effects on vegetation and on animal life in lakes and streams.

## Problems

1. Assume you are going to construct a building from unreinforced blocks of granite. To what height could you build without crushing the granite block at the base of the structure? (1 $in^3$ of granite weighs 0.1 lbs)
2. List some factors that contribute to the range in compressive and tensile strengths of the various types of building stones. Briefly explain the way in which each factor affects strength.
3. Suppose you are thinking of opening a quarry to serve the cemetery headstone industry. Which types of rock would be most suitable and why?
4. Do you think an iron-containing limestone would make a good building stone for a monument? Explain your reasoning.
5. One type of foliated metamorphic rock was once widely used as roofing material on private homes. Why do you think it was selected? Why do you think it is not widely used today?
6. Why is the Mohs hardness scale for minerals *not* particularly useful for choosing a construction stone?

7. Look around your campus and identify the types of natural stone you see used for building blocks, facing stone, or decorative purposes. In the space below, list each type of rock, its location on campus, and the use to which it is being put.

| Type of Rock | Location | Use |
|---|---|---|
| 1. | | |
| 2. | | |
| 3. | | |
| 4. | | |
| 5. | | |
| 6. | | |
| 7. | | |
| 8. | | |
| 9. | | |
| 10. | | |

## Further Reading/References


Bates, R. L., and Jackson, J. A., 1982, *Our Modern Stone Age.* Los Altos, California, William Kaufmann, 132 pp.

Hannibal, J. T., and Park, L. E., 1992, A guide to selected sources of information on stone used for buildings, monuments, and works of art: *Journal of Geological Education,* v. 40, pp. 12–24.

Harries-Rees, K., 1991, Dimension stone review. The new "stone age.": *Industrial Minerals,* no. 290, Nov., pp. 43–52.

Kukal, Z., Malina, J., Malinová, R., and Tesarová, H., 1989, *Man and Stone.* Prague, Czechoslovakia, Geological Survey, 315 pp.

Prentice, J. E., 1990, *Geology of Construction Materials.* New York, Chapman & Hall, 202 pp.

Winkler, E. M., 1975, *Stone: Properties, Durability in Man's Environment,* 2nd ed. New York, Springer-Verlag, 230 pp.

Winkler, E. M. (ed.), 1978, *Decay and Preservation of Stone.* Geological Society of America Engineering Geology Case Histories No. 11, 104 pp.

exercise

# EIGHTEEN

# Petroleum and Natural Gas

Petroleum and natural gas supply about 64% of this nation's energy consumption (petroleum 40%, natural gas 24%), a decline from about 80% twenty years ago. The proportion of our petroleum that is imported has risen considerably during the same period, from about one-third to nearly one-half.

Petroleum and natural gas are part of a group of substances called *fossil fuels,* fuel resources that require millions of years to form naturally but are used by humans at rates high enough to exhaust the supply within the next few centuries. Our supply in the United States will be exhausted much sooner because exploration efforts have been going on for a longer time. We have already found our easily discovered petroleum resources, making further discoveries more difficult and more expensive compared to those of many other countries. But in what geologic settings do petroleum and natural gas occur? How do we go about finding deposits located many thousands of feet below the surface?

Petroleum and natural gas are fluids formed largely from the organic tissues of microscopic marine organisms. These organisms live at the ocean surface and are distributed around the world by currents. After they die they settle to the ocean floor and are buried, and their organic matter is converted to liquid and gaseous *hydrocarbons* by processes not fully understood. Hydrocarbons are chemical compounds composed mostly of carbon and hydrogen. Depending on the temperatures reached during and after the conversion, liquid, gas, or both may be formed. Because the original organisms are so small, their skeletons are deposited with other fine-grained materials such as clay-rich muds. Black muds rich in organic matter are the rocks in which petroleum and natural gas form. The fluid hydrocarbons are squeezed from the lithified muds (shales) into porous, permeable sandstones and carbonate rocks that overlie them in the sedimentary rock column (figure 18.1). The hydrocarbon liquid and gas then move through these rocks toward zones of lower pressure until stopped (trapped) by some type of change in geologic conditions.

Five types of traps are responsible for nearly all our petroleum and natural gas reserves: anticlines, faults, salt domes, unconformities, and facies changes (figure 18.2). No matter which type of trap causes the hydrocarbon accumulation, we can recover no more than half the accumulated amount with our current technology; often we can bring only 20–30% to the surface. As a result it is correct to say, as politicians do, that there is more oil still in the ground than has ever been removed. Unfortunately, however, we are unlikely to recover the bulk of this remaining oil because of both the high cost of developing new technology and the immense amount of low-cost oil available overseas.

## *Resource Terminology*

Numerous terms are used in the petroleum industry and the popular press to describe amounts of hydrocarbons and evaluate available fuel resources.

1. Amounts of petroleum are normally measured in barrels (bbl), where one barrel contains 42 gallons. The standard unit of energy used in

**Figure 18.1**

Photograph of liquid petroleum apparently being squeezed from a porous and permeable sandstone by a human hand. Because of the viscosity of petroleum and the small size of most pores in sandstones, the liquid only rarely leaks out of the rock at the apparent rate shown in the photo.

petroleum-industry literature is the British thermal unit (Btu); one gallon of petroleum generates about 138,000 Btu. Lighting a 100-watt light bulb for one hour requires 341.5 Btu. Amounts of natural gas are measured in cubic feet. In terms of Btu energy equivalents, 6,000 $ft^3$ of gas equals one barrel of oil. One thousand cubic feet of gas heats a typical home for one winter day.

2. *Reserves* are the part of a petroleum accumulation that could be economically extracted at the time the estimate is made. But the prices of petroleum and natural gas fluctuate significantly and can change precipitously as international political currents change. For example, before the Arab oil embargo in 1975, the price of a barrel of oil was about $4; by 1980 the price had reached a high of $41. In 1987 a low of $8 was reached and, since then, the price has climbed erratically up to about $20. Imagine the effect this range from $4 to $41 has on reserve estimates! Unless the estimate is based on a specific price per barrel of oil, its value can be grossly in error.
3. *Secondary Recovery.* Petroleum and natural gas are like artesian water in that the fluid rises into the drill hole because of pressure release. The oil might rise all the way to the surface, forming a gusher, or might require pumping part of the way—neither of which requires pumping anything into the well. In secondary recovery, however, oil is forced from the reservoir rock by pumping water down the hole. This process is also called *enhanced recovery.*
4. *Tertiary recovery* refers to forcing oil from the reservoir pores by methods more exotic than water injection. Examples include the injection of steam, carbon dioxide gas, detergents, or bacteria. As the price of petroleum increases, the use of these exotic procedures becomes economically feasible.

## Oil Pollution

Petroleum from many sources has been entering the ocean for at least a billion years; it is not just a twentieth-century phenomenon. The present rate of natural seepage from the sea floor totals millions of tons per year (one ton equals about 250 gal of petroleum). Even so, the oceans are not covered with a sheet or even a sheen of oil, nor are beaches heavily covered, although many beaches have small quantities of oil. Clearly, natural processes must make the oil that seeps into the oceans disappear fairly rapidly. These processes include evaporation, oxidation, bacterial degradation, and dispersal by winds, currents, and tides.

Currently, however, our civilization faces a problem: the amount of oil added to the ocean is increasing steadily because of the increasing use of large tanker ships to transport imported oil to major consumers, particularly the United States and western Europe (figure 18.3). Tankers now reach lengths of 1,000 ft and hold up to 2 million barrels of oil (84 million gal) so that even a single spill is likely to be a major disaster. This situation has led to considerable study of both the effects of oil pollution on the marine environment and clean-up methods for open-sea and nearshore spills. Current clean-up methods include using booms to control the lateral spread of the spill and "vacuum cleaners" and adsorbents to remove the oil. Burning, sinking, and dispersing spills have also been attempted, but with mixed results. In extreme cases of beach pollution, people have been employed to clean gravel, stone by stone, along miles of beach. Such extreme cases highlight the need for developing new, automated methods for dealing with the tanker spills that are inevitable in our industrial civilization.

**Figure 18.2**

Major types of subsurface traps for petroleum and natural gas. (a) anticline; (b) fault; (c) salt dome; (d) unconformity; (e) facies change or pinch-out. Each trap has a porous and permeable reservoir rock with an impermeable barrier above it. In (c) the barrier is a finger of salt that has risen from a salt layer below; in the other cross-sections, the barrier is a shale unit. The petroleum is overlain by a less dense gas accumulation. Below each petroleum zone is salt water, which is denser than the petroleum.

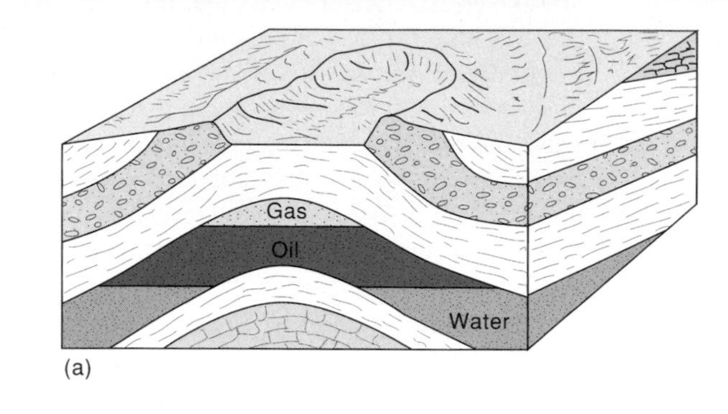


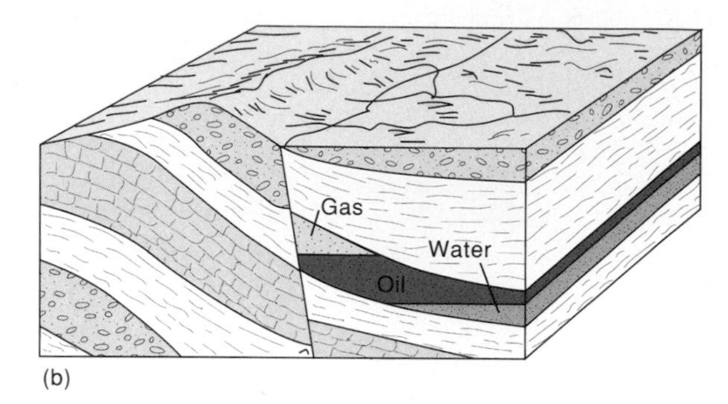


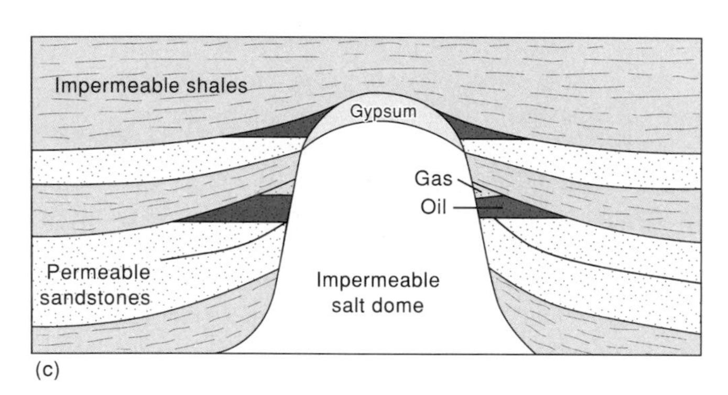


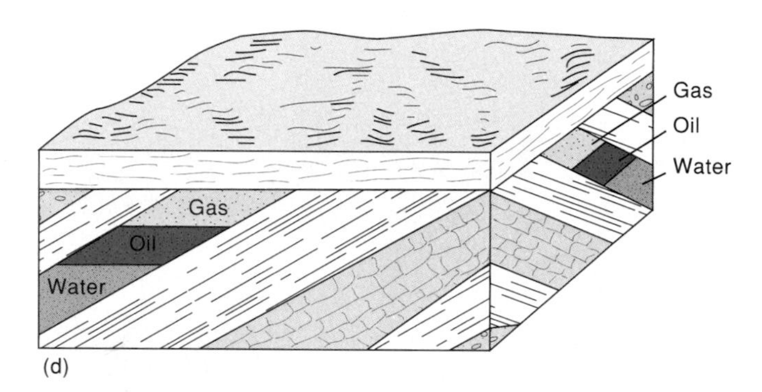

## Problems

1. Some faults are excellent structural traps for oil and/or gas while others are not. Explain why such differences can occur. How might a "leaky" fault help oil-exploration efforts?
2. Explain how the texture and mineral composition of a sand or carbonate sediment affect its ability to serve as a future reservoir for oil and gas.
3. Figure 18.4 shows a township in the Public Land Survey System, with elevations above sea level given for points at the top of a porous, permeable sandstone unit. The Student Petroleum Exploration Club has drilled the 38 holes and discovered 11 producers, shown as black dots. The 27 dry holes are marked with the standard symbol.
   a. Using a 100 ft contour interval, construct a structure contour map of the top of the producing sandstone.
   b. What type of structural trap is present?

**Figure 18.3**

Major international movements of petroleum by ship and major tanker spills between 1974 and 1983.
*Wardley-Smith, J.* Control of Oil Pollution, *published by Graham and Trotman Ltd. Reprinted by permission.*

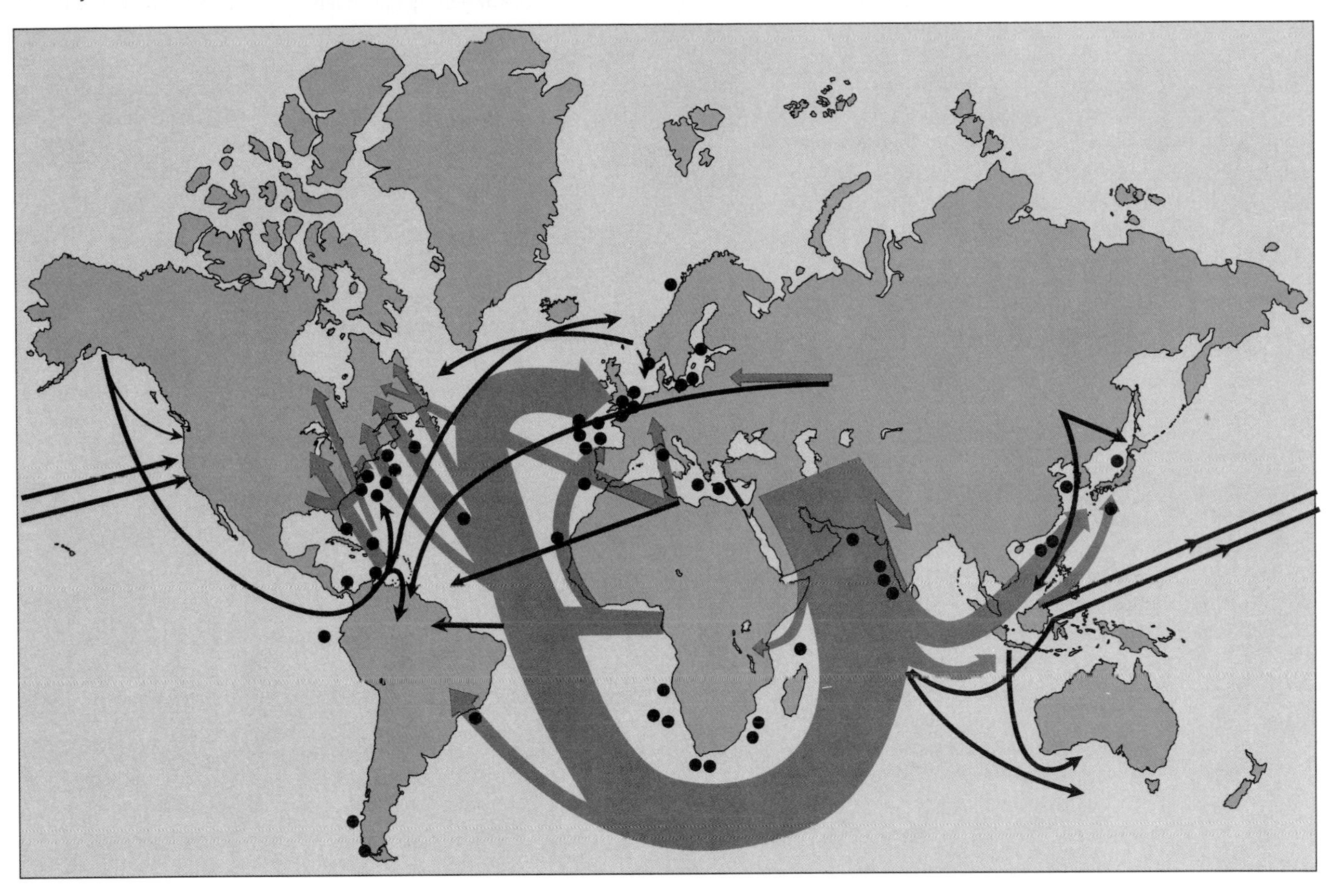

**Figure 18.4**

Elevation (above sea level) of the top of a porous and permeable sandstone.

| | | | | | |
|---|---|---|---|---|---|
| 3270 | 3200<br>3110 | | 3000 | 2880 | 2750 |
| 3150 | | 2950 | | 3130 | 2850<br>3000 |
| | 2980 | 3060 | 3120 | 3370<br>3490 | 3250 |
| 3050<br>3000 | | 3470<br>3430 | 3550<br>3530<br>3400 | 3250 | 3100 |
| 3250 | 3350 | 3370<br>3270 | 3290 | 3230 | 2970 |
| 3100 | 3050 | 2950 | 3050 | 2950 | |

c. How thick do you estimate the oil-producing zone to be?

d. If the sandstone is 500 ft thick and has a porosity of 15%, how many barrels of oil might the reservoir contain?

e. There is some acreage for lease in the center of section 15. Do you think it is a good bet to lease this property and drill? Explain your decision. How about property in the center of section 16?

4. A very large proportion of the actual and potential oil reserves in the United States lie either in Alaska or on the continental shelf. Members of the petroleum industry want to lease these areas for oil and gas exploration and, they hope, production. Environmental groups almost always oppose such efforts because of possible spills and damage to scenery, plants, and animals. How do you believe such disputes should be settled? Is it possible to satisfy both sides?

## Further Reading/References

Baker, R., 1982, *A Primer of Oilwell Drilling,* 4th ed. Austin, Petroleum Extension Service, University of Texas at Austin, 94 pp.

Cole, H. A. (ed.), 1975, *Petroleum and the Continental Shelf of North-west Europe,* v. 2, *Environmental Protection.* New York, John Wiley & Sons, 126 pp.

Gerding, M. (ed.), 1986, *Fundamentals of Petroleum,* 3rd ed. Austin, Petroleum Extension Service, University of Texas at Austin, 452 pp.

Leecraft, J., 1983, *A Dictionary of Petroleum Terms,* 3rd ed. Austin, Petroleum Extension Service, University of Texas at Austin, 177 pp.

Morris, J., House, R., and McCann-Baker, A., 1985, *Practical Petroleum Geology.* Austin, Petroleum Extension Service, University of Texas at Austin, 234 pp.

Selley, R. C., 1985, *Elements of Petroleum Geology.* New York, W. H. Freeman, 449 pp.

Wardley-Smith, J., 1983, *The Control of Oil Pollution.* London, Graham & Trotman, 285 pp.

exercise

# NINETEEN

## Coal

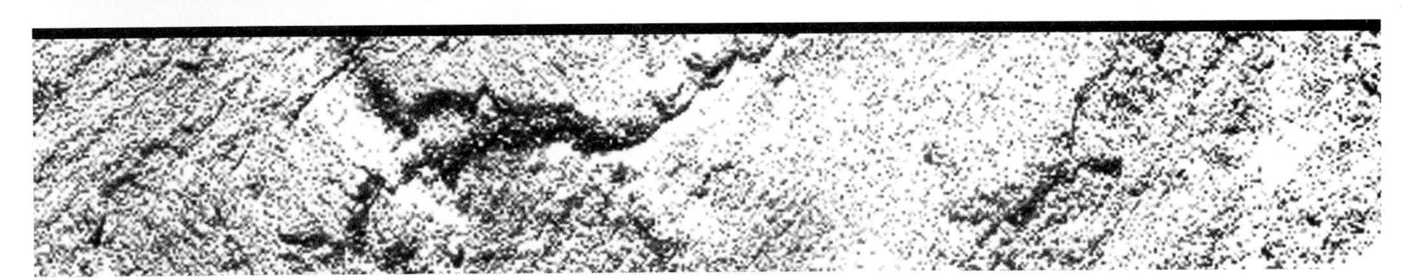

Coal ranks behind oil and gas as a supplier of energy to the American public, accounting for about 24% of our current needs. This percentage has doubled over the past 20 years, despite a 20% growth in total energy consumed. The amount of coal mined in this country has increased dramatically in the past two decades, in part because this fossil fuel is so abundant. About one-fifth of the world's coal reserves are located in the United States (figure 19.1)—sufficient coal to meet our needs for several hundred years.

Coal deposits are much easier to locate than petroleum and natural-gas deposits because large coal deposits almost always form in swamps. Because coal is a solid, the search is straightforward. If we can find ancient swamp environments in the geologic record, we stand a good chance of finding commercially valuable coal reserves. The reason for the association between swamps and coal is that coal forms mostly from partially decomposed remains of land plants. The most luxuriant plant growth occurs in swamps, humid areas where the groundwater table intersects relatively flat ground surfaces that have few throughflowing streams. Many modern coal-forming areas exist on the coastal plain of the eastern United States, from the Carolinas southward into Florida (e.g., Okefenokee Swamp in Georgia). The economically valuable coal reserves found in ancient rocks did not start forming until about 400 million years ago (Devonian Period), when large, multicellular plants first colonized the land surface.

### *Sulfur Content*

As coal forms, the partially decomposed remains of plants pass through several distinctive stages during their transformation to a black, layered, combustible rock (table 19.1). As the dead plants "mature" toward coal at increasing burial depths and, therefore, at increasing temperature and pressure, their heat output when burned rises, as does the percentage of carbon they contain. Thus, a high rank of coal gives off more heat and leaves less ash residue, which can contain toxic heavy metals. These characteristics are, of course, of interest to any homeowners who burn coal. Most coal, however, is used by industries and electric utilities, for whom sulfur content is even more important. The sulfur is emitted from the smokestack as sulfur dioxide gas ($SO_2$), which dissolves in atmospheric moisture to become sulfuric acid, a major constituent of acid rain:

$$2SO_2 + O_2 + 2H_2O \rightarrow 2H_2SO_4$$

Plants, the parent material of coal, average less than 1% sulfur, but this percentage can increase greatly during the early stages of coal formation. The enrichment occurs when the swamp is transgressed by the sea during a time of either rising sea level or coastal sinking. Seawater contains abundant sulfur in the form of sulfate ions (2,650 ppm). As the seawater percolates downward through the plant/peat layers, bacterial activity chemically reduces the sulfate to sulfur ions ($S^{-2}$); the ions then combine with available reduced

**Figure 19.1**

Coal fields of the United States.

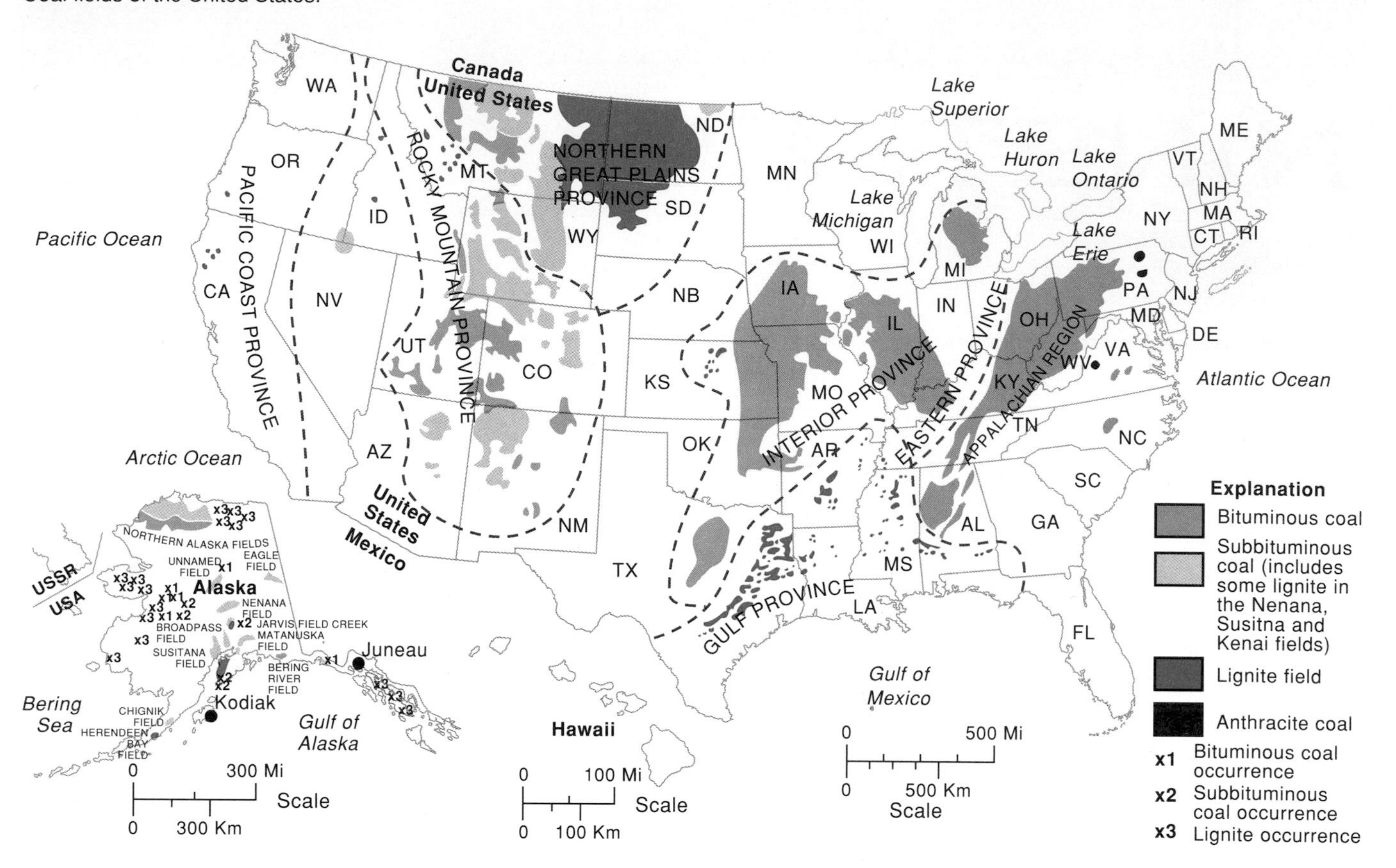


TABLE 19.1

Some Important Characteristics of Organic Material during the Transformation from Living Plants to Coal, Based on Analyses of United States Coal Resources

| Substance | Carbon % (excluding moisture) | Moisture + Volatile Matter (%) | Btu per Pound |
|---|---|---|---|
| Living plant | 50 | — | — |
| Peat | 50–60 | >75 | — |
| Lignite | 60–70 | 75 | <9,000 |
| Subbituminous coal | 70–78 | 70 | 9,000–12,200 |
| Bituminous coal | 78–90 | 60 | 12,200–15,500 |
| Anthracite coal | 92–98 | 8 | ~15,000 |

**Figure 19.2**

Estimates of 1987 recoverable coal reserves by region and sulfur content.

*Source: Energy Information Administration, 1989, p. ix.*

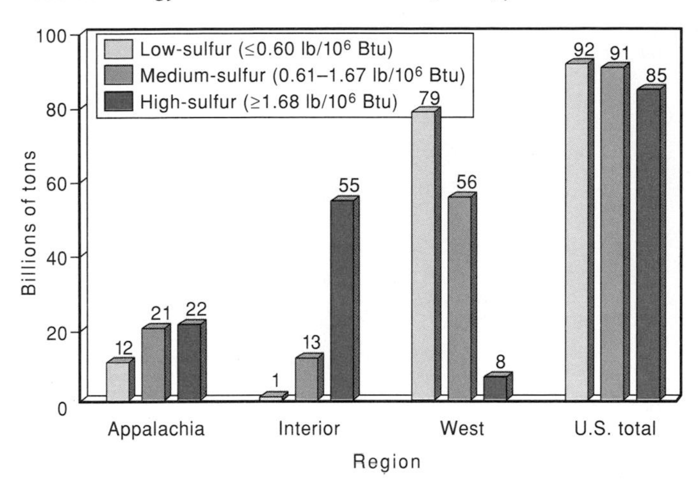


iron ($Fe^{+2}$) to precipitate as pyrite ($FeS_2$). Most of the sulfur in high-sulfur coals occurs as microscopic-size pyrite grains, not as organic sulfur. The sulfur content of U.S. coals ranges from 0.2 to 7%, averaging 1 to 2%.

Because of its cause, sulfur content is unrelated to coal rank (figure 19.2). Western United States reserves contain most of our lower-grade combustible rock, lignite and sub-bituminous coal, and also contain the highest proportion of low-sulfur coal. But reserves in Appalachia and the Interior Province, all of which contain almost entirely bituminous coal, vary greatly in their proportions of low- and high-sulfur types. This variation has considerable economic and environmental significance, because the densely populated East Coast is closest to Appalachian coal, 40% of which has a high sulfur content and must either be treated before burning or "scrubbed" in the exhaust stacks after burning to prevent sulfur emissions. Both treatments increase the cost of the coal but may still be more economical than shipping low-sulfur coal long distances.

## *Coal Mining and the Environment*

Little coal in the stratigraphic column crops out at the surface; most must be mined. As a rule of thumb, coal beds to depths of 200 ft are considered surface-minable (strip mines), while beds 200 to 1,000 ft deep are considered underground-minable. Actual mining decisions also depend on the thickness of the coal seam, the quality and rank of the coal, the difficulty of removing overlying rocks, transportation costs, and the prevailing price of the coal. About two-thirds of the coal mined in the United States is strip-mined. Coal mining, particularly strip-mining, creates both topographic and aesthetic problems. Blasting and use of bulldozers and other equipment removes vegetation and fragments the overburden, greatly increasing both the erosion rate and the sediment load of nearby streams. The sediment can fill streambeds and increase the frequency of flooding. Such problems can be alleviated by rapid land reclamation, now required by federal law. The distressed land must be restored as closely as possible to its pre-mining condition. Smoothing out artificially chopped-up topography and replanting vegetation are costly steps necessary to prevent major environmental damage and aesthetic deterioration.

Underground coal mining also creates environmental problems. Acid mine drainage occurs in many mining localities. In areas of older abandoned mines (figure 19.3), surface structures sometimes collapse into unsupported, shallow underground mines. With increasing age, supporting structures within mines decay, while subsurface waters weaken the rocks through weathering. The subsidence typically occurs as pits or trenches that can be deeper than the thickness of the coal mined. A related and perhaps even more serious problem in old, abandoned mines is spontaneous combustion of the remaining coal. In the mid-1980s the U.S. Bureau of Mines estimated that more than 250 uncontrolled mine fires were burning in 17 states. Such fires can release noxious sulfurous fumes and increase the frequency of surface-structure collapse. Some fires in Pennsylvania have been burning for 25 years, travelling several kilometers during that time. Periodically, both people and property have fallen into newly collapsing pits at the surface. Repeated attempts to extinguish the fires have failed. If you live in a coal-mining area, is your home insured against such a disaster?

All the damage coal mining causes to property and the natural environment is minor, however, compared to its damage to the miners themselves. Miners, particularly those who work underground, inhale large amounts of coal dust. The dust particles are too small to see, but they have disastrous effects on the lungs, causing chronic bronchitis, emphysema, and, in severe cases, death. The accumulation and retention of coal dust in miners' lungs is directly correlated

## Figure 19.3

Modes of ground subsidence above old, abandoned coal mines.

*R. W. Bruhn and others, 1978, Subsidence over the mined-out coal: American Society of Civil Engineers Spring Convention, Pittsburgh, ASCE Preprint 3293, p. 26–55.*

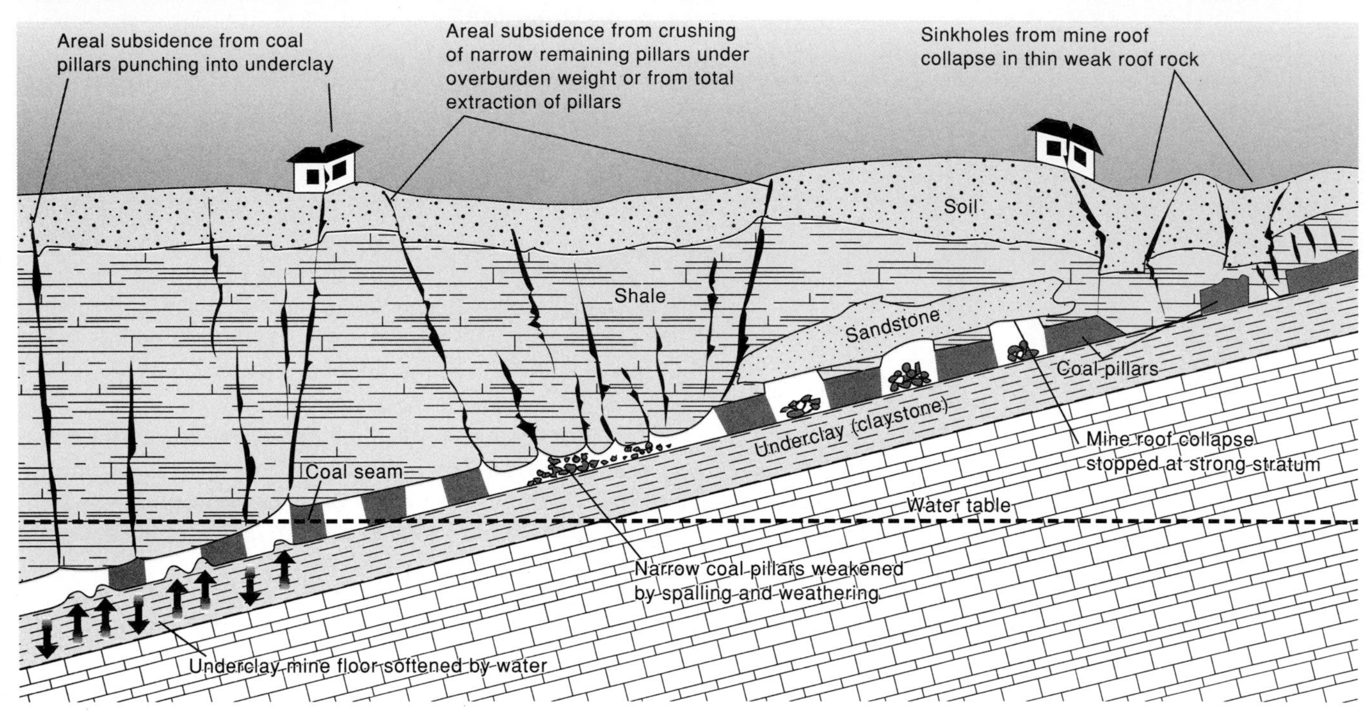


with their years of underground exposure (figure 19.4). Various government studies have found significant coal dust in the lungs of 10 to 30% of current miners and 15 to 75% of ex-miners. Ex-miners have higher percentages partly from spending longer times in underground mines, and partly from the earlier lack of protective measures in the mines. Current measures reduce dust in the mines but cannot eliminate it. Coal mining remains a dangerous occupation.

## Figure 19.4

Relationship between years worked in underground coal mines and incidence of miners' pneumoconiosis.

Archives of Environmental Health, *Volume 27, Page 182, 1973. Reprinted with permission of the Helen Dwight Reed Educational Foundation. Published by Heldref Publications, 1319 Eighteenth St., N.W., Washington, D.C. 20036–1802. Copyright © 1978.*

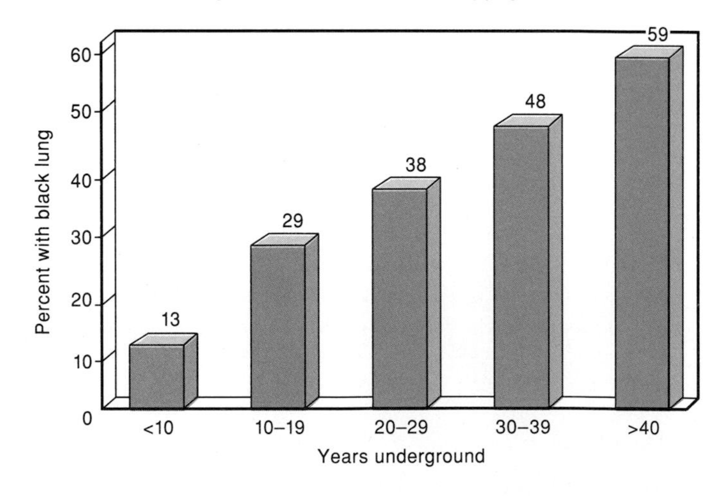


## Problems

1. The Fruitland Formation in northwestern New Mexico and southwestern Colorado is a Cretaceous clastic unit, about $75 \times 10^6$ years old. It is about 300 ft thick and consists mostly of interbedded sandstone, siltstone, and shale, but also contains economically important coal beds in its lower part. Geologic studies have established that when the coals were deposited, the land area was generally to the southwest, while the sea was to the northeast. Figure 19.5 shows the total coal thickness in the Fruitland Formation.
   a. Draw a line through the axis of thickest coal deposits on the map.

**Figure 19.5**

Map showing thickness of coal beds in the Fruitland Formation.

*Source: U.S. Geological Survey.*

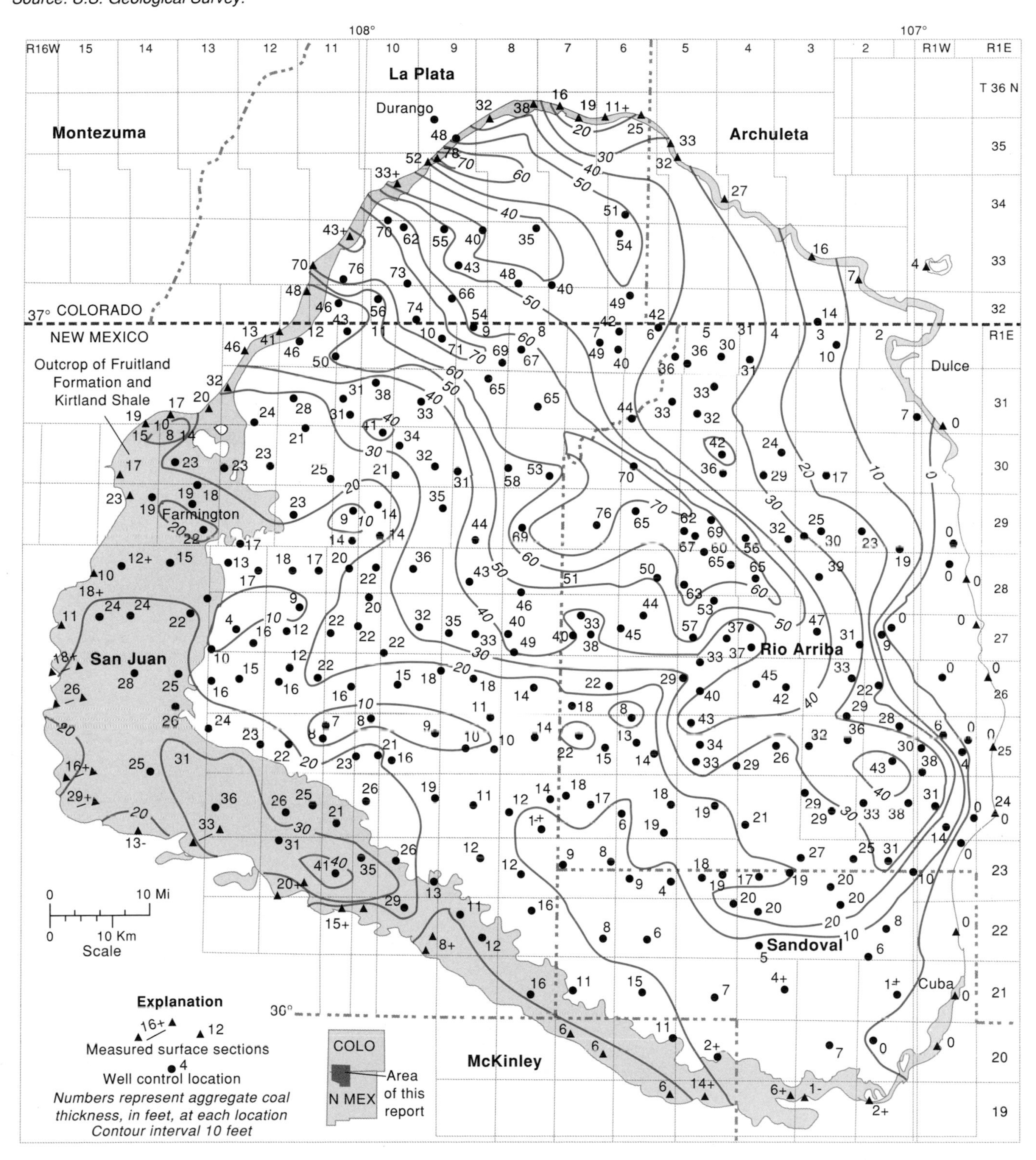

b. What does the orientation of this line tell you about the orientation of the shoreline during the time of coal deposition?
c. Do you think the rate of sinking was the same throughout the entire depositional basin? Why or why not?
d. Most of the coal occurs in the lower one-third of the Fruitland Formation. What does this tell you about drainage conditions during the depositional time of the lower versus the upper Fruitland Formation?
e. Make a numerical estimate of the volume of coal present in the Fruitland Formation.

2. List as many environmental hazards as you can that can result from mining and burning coal.
3. When living organisms die, their tissues normally decay to carbon dioxide plus water, leaving no residue. Why does this not occur in the environment in which most coal forms? (Hint: swamps are stagnant environments.)
4. Considering the requirements for coal formation from dead plant material, examine a geologic map of the United States to explain why all lignite deposits are located in the Montana–North Dakota area rather than in, for example, the Appalachian region.
5. The Carboniferous Period of Earth history (355–290 million years ago) is so named because a major part of the Earth's coal reserves are located in rocks of this age. Considering the depositional setting in which coal precursors live, what might you infer about the latitudinal positions of the Earth's large land masses during this time period? Go to your college library, find a historical geology textbook, and check the latitudinal location of the coal-rich areas of the United States during Carboniferous times.
6. A large percentage of this country's coal reserves are located more than 200 ft underground and cannot be strip-mined. The nation needs access to this coal, and will need it even more in upcoming decades. But because of the dangers of underground mining, these resources are underutilized. Suggest steps that might be taken, either by private industry or by local, state, and federal governments, to alleviate this problem.
7. In Pennsylvania some neighborhoods are settling and houses have collapsed because of underground coal mining. Suppose the coal company is still operating in such an area. What should be done? Are the homeowners entitled to financial relief? From whom? What if the coal company is out of business? What then?

## Further Reading/References


Affolter, R. H., and Stricker, G. D., 1990, *Paleolatitude—A Primary Control on the Sulfur Content of United States Coal.* U.S. Geological Survey Circular 1060, p. 1.

Chadwick, M. J., Highton, N. H., and Lindman, N., 1987, *Environmental Impacts of Coal Mining and Utilization.* New York, Pergamon, 332 pp.

Donaldson, A., et al., 1979, Geologic factors affecting the thickness and quality of Upper Pennsylvanian coals of the Dunkard Basin: in *Carboniferous Coal Short Course and Guidebook,* A. C. Donaldson, M. W. Presley, and J. J. Renton (eds.). Morgantown, West Virginia Geological Survey, p. 133–188.

Energy Information Administration, Office of Coal, Nuclear, Electric and Alternate Fuels, 1989, *Estimation of U.S. Coal Reserves by Coal Type, Heat and Sulfur Content.* Washington, U.S. Department of Energy, 57 pp.

Fettweis, G. B., 1979, *World Coal Resources.* New York, Elsevier, 415 pp.

Kottlowski, F. E., Cross, A. T., and Meyerhoff, A. A., 1978, *Coal Resources of the Americas.* Geological Society of America Special Paper No. 179, 90 pp.

Kroll-Smith, J. S. and Couch, S. R., 1990, *The real disaster is above ground: A mine fire and social comflict.* Univ. of Kentucky Press, Lexington, 210 pp.

Trumbull, J., 1960, *Coal Fields of the United States.* U.S. Geological Survey Map, 1:5,000,000.

Turney, J. E., 1985, *Subsidence above Inactive Coal Mines.* Colorado Geological Survey Special Publication 26, 32 pp.

exercise

# TWENTY

## Radioactive-Waste Disposal

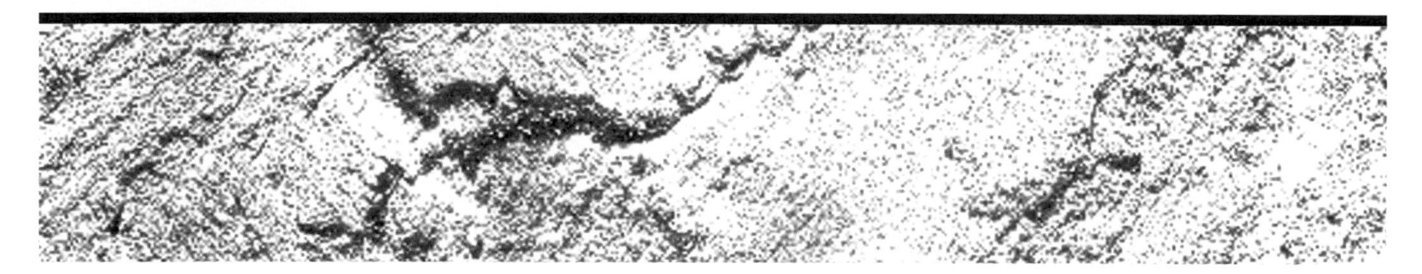

Worldwide, there are currently about 530 nuclear power plants in operation in 35 nations, and both numbers are increasing yearly. These plants all generate very radioactive wastes, or *radwastes,* as does production of nuclear weapons. The U.S. Department of Energy estimates that by the year 2000 the amount of highly radioactive spent fuel and reprocessing waste generated by power plants alone would fill a football field to a height of 254 feet! No method has yet been devised to safely dispose of this growing mountain of material.

Low-level radwaste is perhaps 10 times as abundant as high-level waste and comes from many sources besides reactor operations and reprocessing of spent fuel. Much low-level radwaste comes from the use of radioactive materials in medicine, research laboratories, and industrial processes. Examples of this waste include towels, gloves, carcasses of animals used in biological research, and filter paper.

Radiation arises from the fact that although there are only 88 naturally occurring elements on Earth, each element has one or more isotopes, and most of these isotopes are unstable. Unstable isotopes start to decay or break down from the instant they form (for example, in nuclear power plants). As they decay they give off particles and high-energy radiation that damage living tissues. The proportionate amount of a radioactive isotope that decays in a given time period is a constant. Thus, if we start with 100 g of a radioactive isotope and half of it decays in one year, then half of the remainder (25 g of the 50 g) will decay in the second year, 12.5 g will decay in the third year, and so on. The time required for half of any given amount of an isotope to decay is called the *half-life* of the isotope. In the example, the half-life is one year. Half-lives vary widely from one isotope to another; many isotopes have half-lives of a fraction of a second, while others have half-lives of billions of years (table 20.1). The half-life of an isotope is roughly the inverse of its intensity of radiation, because the greater number of disintegrations in a short-lived material produces stronger radiation. Half-lives are also important because they indicate how long radiation from a given isotope will remain potentially hazardous. As a rule of thumb, in most circumstances the amount of an isotope remaining after 10 half-lives is too small to pose a serious threat to the surroundings.

### *Types and Effects of Radiation*

Three types of radiation are emitted from radioactive substances: *alpha, beta,* and *gamma.* The three differ in both their form and their penetrating power. Alpha rays are particles composed of two protons and two neutrons (helium nuclei) expelled from the nucleus of the radioactive isotope. Beta rays are particles composed of expelled electrons. Gamma rays, in contrast, are a form of very short-wavelength radiation, even shorter than X-rays. Radiation poses a health hazard because it can penetrate human tissue and ionize the atoms in living cells. These ionized atoms have altered electrical charges and therefore different chemical behavior, which upsets normal body chemistry. If the radiation dose is severe enough, radiation sickness results.

**Figure 20.1**

Variation in natural background radiation in the conterminous United States (millirem per year).

*Source: A. W. Klement, Jr., 1972, ORP/CDS 72-1, Environmental Protection Agency, Estimates of ionizing radiation doses in the United States, 1960–2000.*

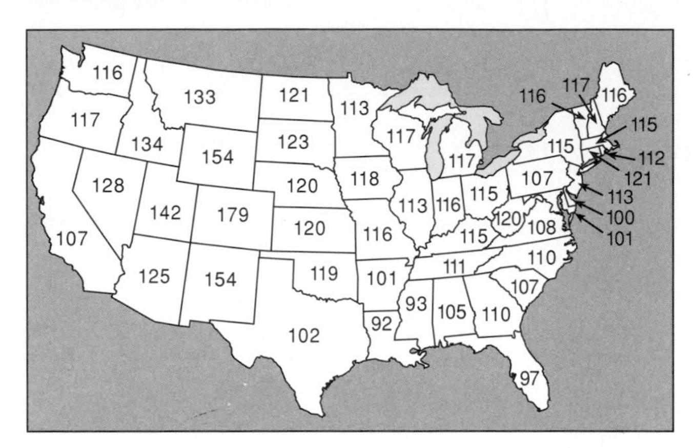


TABLE 20.1

Radioisotopes Important in Waste Disposal

| Isotope | Symbol | Half-Life |
|---|---|---|
| Protoactinium-234 | $^{234}Pa$ | 6.8 hr |
| Radon-222 | $^{222}Rn$ | 3.82 day |
| Bismuth-210 | $^{210}Bi$ | 5.0 day |
| Xenon-133 | $^{133}Xe$ | 5.25 day |
| Polonium-210 | $^{210}Po$ | 138.4 day |
| Cobalt-60 | $^{60}Co$ | 5.27 yr |
| Krypton-85 | $^{85}Kr$ | 10.7 yr |
| Tritium | $^{3}H$ | 12.33 yr |
| Lead-210 | $^{210}Pb$ | 22.3 yr |
| Strontium-90 | $^{90}Sr$ | 28.8 yr |
| Cesium-137 | $^{137}Cs$ | 30.17 yr |
| Americium-241 | $^{241}Am$ | 433 yr |
| Radium-226 | $^{226}Ra$ | 1,630 yr |
| Radiocarbon | $^{14}C$ | 5,730 yr |
| Plutonium-240 | $^{240}Pu$ | 6,570 yr |
| Americium-243 | $^{243}Am$ | 7,370 yr |
| Curium-245 | $^{245}Cm$ | 8,500 yr |
| Plutonium-239 | $^{239}Pu$ | 24,000 yr |
| Thorium-230 | $^{230}Th$ | 80,000 yr |
| Tin-126 | $^{126}Sn$ | ca. $10^5$ yr |
| Technetium-99 | $^{99}Tc$ | $2.14 \times 10^5$ yr |
| Uranium-234 | $^{234}U$ | $2.45 \times 10^5$ yr |
| Neptunium-237 | $^{237}Np$ | $2.14 \times 10^6$ yr |
| Cesium-135 | $^{135}Cs$ | $3 \times 10^6$ yr |
| Iodine-129 | $^{129}I$ | $1.6 \times 10^7$ yr |
| Uranium-235 | $^{235}U$ | $7.04 \times 10^8$ yr |
| Uranium-238 | $^{238}U$ | $4.47 \times 10^9$ yr |

Effects of radiation are generally measured in *rads,* short for *r*adiation *a*bsorbed *d*ose; a rad is the amount of radiation required to liberate $10^{-5}$ joules of energy per gram of absorbing material. To help define the actual effects of such radiation on humans, a second unit of measurement known as the *r*oentgen *e*quivalent *m*an or *rem* is used, with one rem equal to one rad of X-rays or gamma rays. All radioactivity is considered damaging, but its specific effects depend on the activity, distance, and shielding of the source, as well as the duration of exposure, the type of radiation, and the type of tissue irradiated. The average annual whole-body dose due to natural radiation from rocks, cosmic rays, and other sources is about 0.1 rem/yr (figure 20.1). Medical X-rays and color television may increase the annual dose to 0.2 rem. The known effects of high doses of radiation are shown in table 20.2. Low-energy alpha rays are not very penetrating and can be stopped by a sheet of paper or by the outer layer of human skin (though the long-term result might be skin cancer). Higher-energy beta rays are more penetrating and can pass through about one inch of human flesh or water but are stopped by a thin sheet of aluminum. High-energy gamma rays are the most penetrating and damaging but can be stopped by dense materials such as concrete and lead.

## *Waste-Disposal Sites*

Radioactive waste should be isolated for at least 10 times the half-life of the longest-lived, dominant radioisotopes; the half-lives of the principal components of low-level radioactive waste are less than 30 years. Thus, waste containing primarily strontium-90 and cesium-137 requires 300 years of isolation (see table 20.1). Waste dominated by radioisotopes with half-lives greater than that of cesium are considered high-level waste and have somewhat different isolation requirements. For both types of waste the key concern is to keep the radioisotopes from entering the water supply, as has occurred at some disposal sites (figure 20.2). For this reason, current scientific opinion favors disposal of low-level radwaste in excavations above the groundwater table in arid or semiarid areas, such as those in the western and southwestern United States. These disposal sites offer several advantages: the water table is hundreds of feet below the ground surface and very unlikely to rise significantly during the next 300–500 years; little water percolates downward in such dry regions; cavities tens of feet deep can be excavated easily; monitoring of excavated storage

TABLE 20.2

Effects of Whole-Body Radiation

| Dose (rems) | Effects |
|---|---|
| 0–25 | A dose around 25 rem may reduce the white blood cell count |
| 25–100 | Nausea for about half those exposed, fatigue, changes to blood |
| 100–200 | Nausea, vomiting, fatigue, death possible, susceptible to infection (low white blood cell count) |
| 200–400 | A lethal dose for 50% of those exposed, especially in absence of treatment. Bone marrow, spleen (blood-forming organs) damaged |
| >600 | Fatal, probably even with treatment |

*From J. E. Ferguson,* Inorganic Chemistry and the Earth. *Copyright © 1982 Pergamon Press, Oxford, England. Reprinted by permission.*

cavities is simple; and backfill from the cavities can be removed easily should problems develop tens or hundreds of years from now. The natural advantages of these disposal sites could be further improved by barriers engineered to prevent water infiltration.

High-level radwaste poses more difficult problems because of the long half-lives of the isotopes (table 20.1). This waste is too dangerous to be stored above ground for extended periods; other methods, such as disposal in space, in a glacier, or on the ocean floor are either too dangerous or too expensive. At present the favored disposal method for high-level radwaste is underground storage. Current plans call for filling numerous canisters with hot, highly radioactive material, putting them deep underground, and keeping them there in isolation for at least 10,000 years. But how will such prolonged exposure to heat and radiation affect the rocks enclosing the repository? Can the canisters be kept from contact with water indefinitely? Will they corrode? How fast will the radioisotopes dissolve and possibly move out of the containment area? Currently we have no reliable answers to such questions, but research is continuing.

Burial sites for high-level radwaste have clear geologic requirements. The rock at the appropriate depth must be strong enough to maintain an opening, should have low permeability and few fractures through which groundwater might move, and should not be in an area prone to earthquakes, volcanic activity, or severe erosion. Rocks best suited to radwaste storage should also have high heat conductivity to help keep temperatures low. The rock types that best satisfy all these criteria are bedded salt or salt domes, granite, basalt, argillaceous rocks, and tuffaceous rocks.

The scientific problems concerning radwaste disposal are serious, but even more difficult are the political problems. Most people are unwilling to live near a disposal site. Americans fear radioactivity so strongly that no assurances by scientific or governmental offices seem adequate. Almost everyone agrees that radwaste disposal is necessary, but few people want a disposal site nearby. Part of the fear stems from the invisibility of radiation and the natural fear of unseen things, and part from a distrust of governmental pronouncements. The federal government has often underestimated pollution dangers to the public, and people have become understandably wary. But radioactive waste must be placed somewhere; allowing it to remain above ground is even more dangerous than storing it below ground. The Environmental Protection Agency has an important education job to do on this issue.

## Problems

1. Figure 20.1 shows that the highest values of natural background radiation occur in the western states (128–179 millirem), and the lowest values occur along the Gulf and Atlantic coasts. Suggest two likely explanations for these patterns. (Hint: Examine a geologic map and an elevation map of the United States.)
2. Suppose a container of radwaste holds 130 g of curium-245. How long will it take for the amount present to reduce to 1 g?
3. During the period of above-ground nuclear bomb testing in the United States (1946–1963), abundant tritium entered the atmosphere and appeared in rainwater. Suggest ways in which this fact might be useful for studying any part of the hydrologic cycle.
4. Figure 20.2 shows the result of poor disposal practices in the 1940s at a site near Chicago, Illinois.
   a. What is the approximate thickness of the glacial drift?
   b. What do you think controls the thickness of the "variably saturated zone"?
   c. Contour the tritium values at a logarithmic contour interval of powers of ten (10, 100, 1,000, etc.).
   d. The contours of tritium (radioactive $^3H$) values do not form a concentric hemispherical pattern; instead, they form elongated fingers. What does this pattern indicate about the internal structure of the glacial drift?
   e. The contours appear compressed at the boundary between the drift and the underlying dolostone. What might explain this pattern?
   f. The dolostone contains both horizontal and vertical joints. What effect might these have on tritium migration?
   g. Presumably the people who live near Plot M are not overjoyed about the high tritium levels in their groundwater supply. As County Supervisor,

## Figure 20.2

Cross-section showing the vertical distribution of tritium concentrations in groundwater near burial site Plot M near Chicago, Illinois, in 1981 and 1983. The low-level radwaste was buried from 1943 to 1949. Background tritium concentration is 0.2 nCi/L. Vertical lines are wells from which data were obtained.

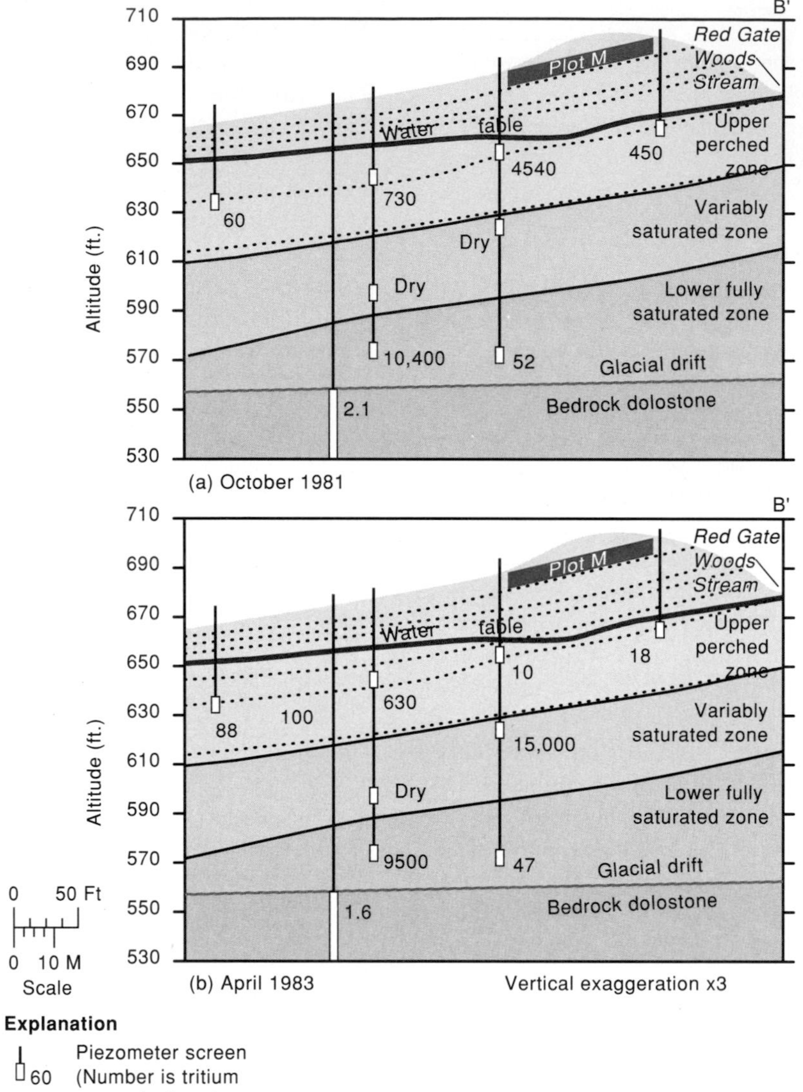


you receive many complaints about the real or imagined dangers of imbibing tritium in amounts thousands of times the background amount. How do you deal with these complaints?

h. As a resident of the area, should you consider relocating? In which direction and how far should you move? (You work in Chicago about 20 miles northeast of your home.)

5. You are interested in the possible effects of natural radiation on the frequency of birth defects.
   a. Design a program of data collection to study this question.
   b. Suppose your investigation reveals that pregnant women in certain parts of your state are twice as likely to give birth to children with birth defects as women in other parts of the state. What do you think the state government should do with this information? Do your results have any financial implications for the state's population?
   c. Suppose a nationwide study of the type you have made indicates that stillbirths and birth defects are twice as frequent in Colorado as in neighboring Nebraska. Should the federal government develop a policy to deal with this "unfair" variation among states? If so, what?

## Further Reading/References

Battelle Memorial Institute, Office of Nuclear Waste Isolation, 1982, *Answers to Your Questions about High-level Nuclear Waste Isolation.* Columbus, Ohio, 69 pp.

Bedinger, M. S., 1989, *Geohydrologic Aspects for Siting and Design of Low-level Radioactive-Waste Disposal.* U.S. Geological Survey Circular 1034, 36 pp.

Chapman, N. A., and McKinley, I. G., 1987, *The Geological Disposal of Nuclear Waste.* New York, John Wiley & Sons, 279 pp.

Fischer, J. N., 1986, *Hydrogeologic Factors in the Selection of Shallow Land Burial Sites for the Disposal of Low-level Radioactive Waste.* U.S. Geological Survey Circular 973, 22 pp.

Krauskopf, K. B., 1988, *Radioactive Waste Disposal and Geology.* New York, Chapman and Hall, 145 pp.

La Sala, A. M., Jr., et al., 1985, *Radioactive Waste. Issues and Answers.* Arvada, Colorado, American Institute of Professional Geologists, 27 pp.

Nicholas, J. R., and Healy, R. W., 1988, *Tritium Migration from a Low-level Radioactive-Waste Disposal Site near Chicago, Illinois.* U.S. Geological Survey Water-Supply Paper 2333, 46 pp.

Roxburgh, I. S., 1987, *Geology of High-level Nuclear Waste Disposal.* New York, Chapman and Hall, 229 pp.

# exercise TWENTY-ONE

## Surface-Water Pollution

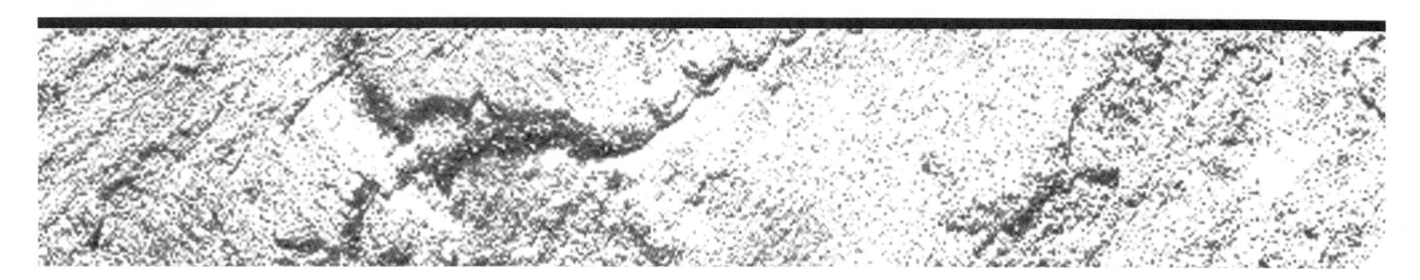

The volume of surface water on the Earth is limited; existing supplies must be carefully managed in terms of both conservation and cleanliness. Both considerations are necessary if the world is to avoid a catastrophe of unimaginable proportions. Conservation can be achieved through public education, but keeping our water supplies clean is a much more difficult problem. Unless more effort is expended on keeping our water pure, we might someday need to boil, filter, and decontaminate the water in our homes before we use it. Such a prospect sounds like science fiction, but it will become all too real if the present trend toward increasing pollution continues.

Pollutants or contaminants can consist of any of the following, each of which creates different problems for treatment:

1. Microorganisms, including pathogenic viruses and bacteria
2. Organic matter, primarily from domestic urban sewage and from rural septic tanks
3. Chemical wastes from industrial operations such as mining and petroleum exploration; from leaks in pipelines and storage tanks; from agricultural operations, such as runoff from animal feedlots and fields treated with pesticides and fertilizers; and from leaching of not-so-sanitary landfills
4. Nuclear wastes generated by power plants, weapons manufacturing plants, laboratories, and medical research facilities

### *Water-Quality Standards*

Requirements for water purity depend on the use to which the water is to be put. Drinking requires the highest purity; irrigation has lower standards, at least for many substances. The federal government has developed criteria for all categories of water quality, including bacterial content, physical characteristics, and chemical constituents. Normally problems involving bacterial content or physical characteristics can be alleviated. Removing or neutralizing undesirable chemical contaminants, in contrast, is often both difficult and expensive. The presence of chemical impurities often imposes major limitations to water utilization. Removal of many organic pollutants poses problems that are as yet unsolved.

In addition to difficulties involving measurable contaminants, we face the rarely mentioned problem of substances present in amounts too small to detect with current equipment. For example, for some pollutants the detection limit may be in parts per million but the substance may be harmful in amounts as low as parts per billion. A related, rarely discussed problem is that we know so little about the effects of many substances on humans that we have set no drinking-water standards for those substances. Without a perfect knowledge of human biochemistry, which can never be achieved, we can never be certain which substances in which amounts endanger human health. Unless we completely dismantle our industrial civilization and return to living in caves, however, we will always be poisoning ourselves to a greater or lesser extent. We cannot solve the pollution problem completely; instead, we must do the best we can with our current tools and understanding.

## *Residence Time*

When we find that an area or aquifer is polluted, we must answer three questions:

1. What is the source of the pollutant? Sources can be of two types: point sources such as a septic tank, an oil spill, or an industrial waste outlet; or nonpoint sources such as farmland runoff or strip-mine drainage. Point sources are much easier to find and remedy.
2. What is the nature of the pollutant? Is it a single element such as arsenic, cadmium, or lead? A complex organic compound such as PCBs (polychlorinated biphenyls) or vinyl chloride? A mineral group such as asbestos? A radioactive by-product from an industrial or military project? Each of these materials poses different treatment problems.
3. What are the migration pattern and expected lifetime of the pollutant? Pollutants can be solids, liquids, or gases. Many are destroyed quickly in the natural environment, perhaps by interaction with bacteria. Radioactive materials can self-destruct to an acceptably low level within a few days or can linger for millions of years. But the lifetimes of many industrial and chemical pollutants are unknown and, for safety reasons, must be assumed to be very long.

When we evaluate a polluted aquifer, we often express the duration of the pollutant as its residence time, defined as

$$R = \frac{C}{F}$$

where $R$ is the residence time, $C$ the capacity of the reservoir, and $F$ is the rate of inflow and outflow of the compound. Because inflow and outflow rates can change, the residence time is strictly accurate only at the time the measurements are made. Note that this definition of residence time does not consider that the pollutant might decompose or break down into other chemicals before it outflows.

Calculating the residence times of pollutants is fairly easy if we make some simplifying assumptions. If we are dealing with a surface water supply, such as a reservoir, we assume that no water evaporates from the lake surface or infiltrates downward out of the lake. When dealing with a subsurface aquifer, we assume that there is no leakage through the confining beds or infiltration of water upward from other aquifers located below the polluted aquifer. In other words, we assume that the affected body of water is a closed system.

*Case 1.* The amounts of harmful material (e.g., lead-bearing clay, asbestos, PCBs) entering and leaving a reservoir are equal, and the incoming material does not mix with the water in the reservoir. Clearly this assumption is unrealistic, but it does serve as a starting point. Suppose a reservoir containing 200 million gallons of water is fed by a polluted stream inflowing at 20,000 gal/min. Then

$$R = \frac{2 \times 10^8 \text{ gal}}{2 \times 10^4 \text{ gal/min}} = 10^4 \text{ min} = 166.7 \text{ hr} = \sim 7 \text{ days.}$$

Thus, seven days after the polluted water enters the reservoir, it will have flushed out all the pure water and begun to leave. From this point on, all the water in the reservoir is polluted.

*Case 2.* Using the more realistic assumption that the inflowing, polluted water does mix with the unpolluted water in the reservoir, the polluted water would occupy 50% of the reservoir volume after 7 days. In each succeeding 7-day period, half the remaining original reservoir water is replaced. Thus, after 14 days the reservoir contains 75% polluted stream water; after 21 days, 87.5%; after 28 days, 93.75%. The mathematical relationship that describes this decreasing percentage of original reservoir water is $(1/2)^x$, where 1/2 is the ratio and the exponent is the number of 7-day increments. Note that the original, unpolluted reservoir water is never flushed out completely, although after about one month the amount becomes negligible.

*Case 3.* Our reservoir-pollution problem must consider one further complication: the water might be draining from the lake faster than the polluted water is entering it, that is, the level of the lake might be dropping. For example, assume that the 200 million gallons of water in the reservoir is draining out through a dam at 500 gal/min, while the polluted inflow enters at 30 gal/min. How much water will remain in the reservoir after 90 days (129,600 min)?

$$200{,}000{,}000 \text{ gal} + (30 \text{ gal/min} \times 129{,}600 \text{ min}) - (500 \text{ gal/min} \times 129{,}600 \text{ min}) = 139{,}088{,}000 \text{ gal}$$

About 30% of the water has been lost.

What, then, is the residence time of the polluted water remaining in the reservoir at the end of the 90-day period? We can obtain this figure by dividing the amount of water remaining in the reservoir by the net outflow:

$$R = \frac{139{,}088{,}000 \text{ gal}}{470 \text{ gal/min}} = 295{,}932 \text{ min} \approx 206 \text{ days} \approx 7 \text{ months}$$

Real-world situations are more complex than the above examples. The reservoir might have inflow from several polluted streams, for example, each with a different rate of flow. Further complications arise from the fact that streamflow rates vary widely for individual streams. During the dry season streamflow might decrease to zero at the same time that evaporation from the lake surface increases, particularly if the lake has a large surface area. But although we might need to consider more variables, we can still use the principles described to calculate the residence time of pollutants.

## Problems

1. A reservoir contains 2 million gallons of water. A factory that produces pesticides is discharging arsenic into the stream that flows into the reservoir.
   a. The stream flows into the reservoir at 40 gal/min, and the lake drains at the same rate. Assuming that the stream water does not mix with the lake water, how long will it take for the arsenic-laden water to replace the water in the reservoir?
   b. A competing firm located a short distance from the shore of the reservoir has been emptying into the reservoir the residue from manufacturing their anticoagulant rat poison. They have been doing so for some years, and the lake now contains 25% of their contaminated water and 75% uncontaminated water. Assume that there has been no inflow or outflow from the lake and that the anticoagulant has mixed completely with the original, unpolluted reservoir water.

      If the arsenic-laden stream now begins its 40-gal/min inflow, and drainage from the reservoir is 40 gal/min, how long will it take for the amount of anticoagulant pollutant in the reservoir to reduce to 10%? To 1%? Assume the stream water does not mix with the reservoir water.
2. Three streams enter a reservoir that holds 1 million gallons of water. The lake is drained through a dam at a rate equal to the sum of the flow rates of the inflowing streams. Stream A inflows at 10 gal/min and contains 20% PCBs; stream B inflows at 20 gal/min and contains 10% vinyl chloride; stream C inflows at 20 gal/min and contains 5% Red Dye No. 3.

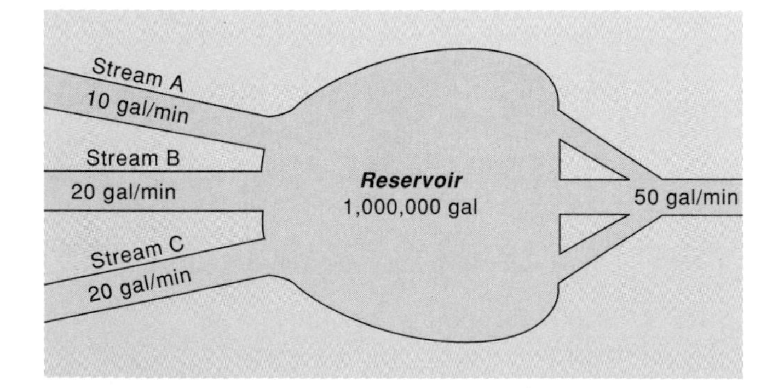


   a. Assuming no mixing, how long will it take for one-half the lake to become polluted?
   b. In the polluted one-half, what percentage of the water is PCBs? What percent is vinyl chloride? What percent is Red Dye No. 3? How many gallons of Red Dye No. 3 are in the reservoir?
   c. When 75% of the original lake water has been displaced, what is the percentage of each pollutant in the lake? How many gallons of each pollutant are in the lake?
   d. Suppose the Environmental Protection Agency decrees an immediate end to further increases in pollution in the lake. If 75% of the original water in the lake was polluted before the edict was issued, how long would it take for each pollutant to decrease to 1%? Assume complete mixing of the pollutants with the water in the lake.
3. Most reservoir bottoms are covered by mud carried in by inflowing streams. How do you think the volume of mud, the percentage of clay minerals in the mud, and the type of clay mineral in the mud would affect the answers to the various questions in problem 2?
4. How do you think the rate of evaporation from the lake surface would affect the answers in problem 2?

## Further Reading/References

Gilliom, R. J., Alexander, R. B., and Smith, R. A., 1985, *Pesticides in the Nation's Rivers, 1975–1980, and Implications for Future Monitoring*. U.S. Geological Survey Water-Supply Paper 2271, 26 pp.

Smith, J. A., Witkowski, P. J., and Fusillo, T. V., 1988, *Manmade Organic Compounds in the Surface Waters of the United States—a Review of Current Understanding*. U.S. Geological Survey Circular 1007, 92 pp.

Smith, R. A., Alexander, R. B., and Wolman, M. G., 1987, *Analysis and Interpretation of Water-Quality Trends in Major U.S. Rivers, 1974–81*. U.S. Geological Survey Water-Supply Paper 2307, 25 pp.

Stamer, J. K., Yorke, T. H., and Pederson, G. L., 1985, *Distribution and Transport of Trace Substances in the Schuylkill River Basin from Berne to Philadelphia, Pennsylvania*. U.S. Geological Survey Water-Supply Paper 2256-A, 45 pp.

Witkowski, P. J., Smith, J. A., Fusillo, T. V., and Chiou, C. T., 1987, *A Review of Surface-Water Sediment Fractions and Their Interactions with Persistent Manmade Organic Compounds*. U.S. Geological Survey Circular 993, 39 pp.

exercise

# TWENTY-TWO

## Groundwater Pollution

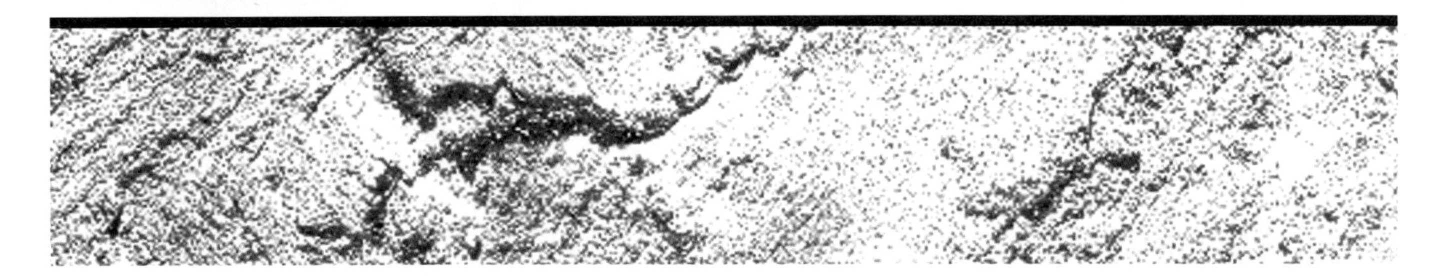

The widespread use of chemical products, coupled with the disposal of large volumes of waste materials, creates the potential for extensive groundwater contamination. Some of the most prominent areas of contamination, such as Love Canal in New York State, have attracted public notice, but for every one of these there are a large number of smaller contamination problems that have, until very recently, gone unnoticed. All students of groundwater-pollution problems agree that episodes such as Love Canal are only the proverbial tip of the iceberg. New instances of groundwater contamination are being identified continually in urban, industrial, and agricultural settings (figure 22.1). Many of these pollution problems have existed for some time but are only now being recognized, thanks to our developing analytical capabilities and increasing concern about the effects of impure water on human health and the environment. Mining activities in the western United States, for example, started during the late 1800s, when Americans worried little about the environment and knew almost nothing about the effects of chemical pollutants on human health. As a result, mining debris was not dealt with properly. Now, however, that same debris is considered hazardous to humans and to other animal and plant life. In some areas of the American West, hazardous levels of elements such as lead, zinc, mercury, or chromium occur in surface or groundwater supplies. Currently applicable Superfund laws (Comprehensive Environmental, Liability, and Compensation Act of 1980, as amended) state that the current owner of the mining property is legally and financially responsible for damage to the water supply and related human health, regardless of whether the mine is still active or whether the present owner had even been born when the pollution occurred.

The pollution problems left over from the last century pale in comparison, however, to those being created by the immense volumes of toxic organic and inorganic materials produced by modern industries, both in the United States and overseas. Some areas of eastern Europe have already been rendered uninhabitable by pollution of the past 50 years. Many industrially produced chemicals are quite stable in groundwater and can pose a serious threat to public health. Current estimates suggest that 0.5 to 2.0% of the groundwater in the conterminous United States is contaminated from point sources. The estimate does not include contamination from nonpoint sources such as agricultural runoff. Certain areas, especially those with a high population density, might have much higher percentages of contaminated groundwater. The EPA has a growing list of more than 400 contaminated sites.

### *Groundwater Contaminants*

The toxic materials that enter groundwater come from many sources (figure 22.2), including septic systems, runoff from agricultural fields and animal feedlots, landfills, accidental leaks and spills, mining debris, ruptures in underground storage tanks such as those under gasoline service stations, decomposing bodies in graveyards, disposable diapers, cheesy pizza sludge, and underground

**Figure 22.1**

Groundwater pollution problems as recognized in the mid-1970s.

*Source: U.S. Water Resources Council, 1978,* The Nation's Water Resources, 1975–2000. *v. 1: Summary, Second National Water Assessment. 86 pp.*

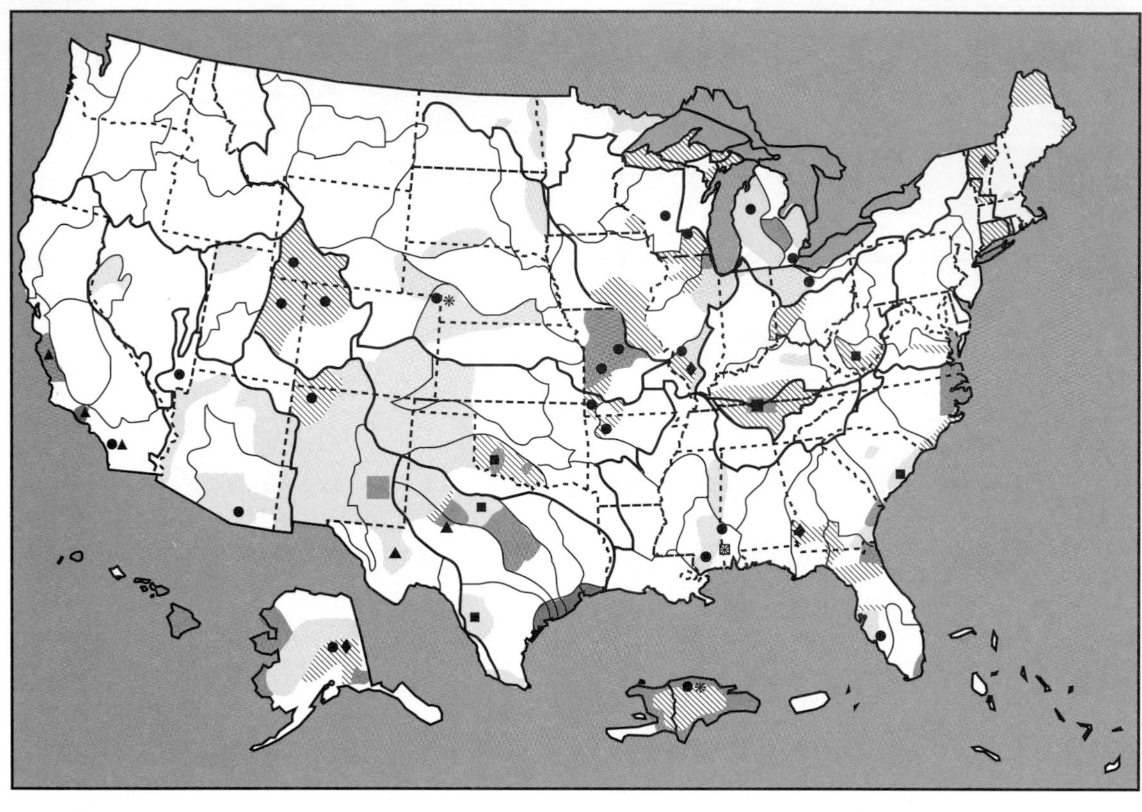

**Area Problems**

- Significant groundwater pollution occurring
- Saltwater intrusion, or groundwater naturally salty
- High level of minerals or other dissolved solids in groundwater
- Unshaded area might not be problem-free, but problem not considered major

**Boundaries**

- Water resources region
- Subregion

**Specific Sources of Pollution**

- Municipal and industrial wastes, including wastes from oil and gas fields
- ● Toxic industrial wastes
- ◆ Landfill leachate
- ▲ Irrigation return waters
- ■ Wastes from well-drilling, harbor dredging, and excavation for drainage systems
- ✻ Well injection of industrial waste liquids

injection of hazardous waste. Given their wide variety of sources, it is not surprising that the types of groundwater contaminants are extremely varied. Some are simple inorganic ions, such as nitrate from fertilizer and feedlot wastes, chloride from deicing salts and saltwater intrusion, and heavy-metal ions from plating works and many other industrial processes. Other contaminants are much more complex synthetic organic compounds that result from industrial and manufacturing processes and the use of pesticides and household cleaners. In many instances we know nothing about the chemical stabilities and lifetimes of the complex compounds, or their effects on human biochemistry.

The magnitude of any pollution problem depends on the size of the area affected; the amount of the pollutant involved; the solubility, toxicity, and density of the pollutant and its persistence in the environment; the mineral composition and hydraulic characteristics of the soils and rocks through which the pollutant moves; and the potential effect of the pollutant on groundwater use. For example, if groundwater contaminants exceed the federal standards for

TABLE 22.1

Maximum Allowable Concentrations of Dissolved Inorganic Substances in Drinking Water According to the Environmental Protection Agency

| Contaminant | Maximum Allowable Level (ppm) |
|---|---|
| Antimony | 0.01 |
| Arsenic | 0.05 |
| Barium | 1.00 |
| Boron* | 1.00 |
| Cadmium | 0.01 |
| Chloride* | 250.00 |
| Chromium | 0.05 |
| Copper* | 1.00 |
| Fluoride | 1.4–2.4 |
| Hydrogen sulfide* | 0.05 |
| Iron* | 0.30 |
| Lead | 0.05 |
| Manganese* | 0.05 |
| Mercury | 0.002 |
| Nitrate (as N) | 10.00 |
| Selenium | 0.01 |
| Silver | 0.05 |
| Sulfate* | 250.00 |
| Zinc* | 5.00 |
| Total Dissolved Solids* | 500.00 |

*Values for substances with asterisks are the EPA-reasonable goals for drinking water but are not federally enforceable. Many of the elements listed are essential for human nutrition in small amounts. Standards also exist for bacteria and for many industrial organic compounds.

drinking water (table 22.1), then the water is considered hazardous to drink. These standards, however, include only a limited number of chemicals and do not protect humans or the environment against all possible contaminants. The long-term effects of even small amounts of many pollutants are unknown.

Given these complexities, we can prevent significant groundwater pollution only by selecting waste-disposal sites in such a way that

1. An adequate thickness of unpolluted sediment containing clay and/or organic material is present both above and below the waste. Both clay and organic matter absorb trace metals and certain complex organic pollutants, preventing these substances from entering the groundwater.

TABLE 22.2

Ranges of Hydraulic Conductivity and Permeability in Sand, Sandstone and Limestone, and Shale

| Material | Hydraulic Conductivity (m/s) | Permeability (darcies) |
|---|---|---|
| Sand | $10^{-2}$ to $10^{-6}$ | $10^{3}$ to $10^{-1}$ |
| Sandstone and limestone | $10^{-3}$ to $10^{-10}$ | $10^{1}$ to $10^{-5}$ |
| Shale | $10^{-9}$ to $10^{-13}$ | $10^{-4}$ to $10^{-8}$ |

2. The disposal areas are as close as possible to places of natural groundwater discharge, so that any pollutant that enters the groundwater will be washed out quickly. Of course, this polluted surface discharge must not be permitted to flow for long distances overland or infiltrate downward into the subsurface.

### *Contaminant Movement*

Groundwater movement in a homogeneous, confined aquifer is determined by three properties:

1. *Hydraulic conductivity,* which includes properties of both the aquifer and the fluid (table 22.2) and is expressed by the equation

$$K = \frac{k\rho_w g}{\mu}$$

$K$ = hydraulic conductivity (cm/s)
$k$ = permeability (darcies = cm$^2$ × $10^{-8}$)
$\rho_w$ = density of water in the aquifer (gm/cm$^3$)
$g$ = acceleration of gravity (cm/s$^2$)
$\mu$ = fluid viscosity (poises, gm/cm s)

2. *Potentiometric gradient,* or slope of the water table, expressed as

$$\frac{\Delta h}{\Delta l}$$

$\Delta h$ = hydraulic head (cm)
$\Delta l$ = distance of flow (cm)

3. *Effective porosity,* the decimal proportion of interconnected pores, expressed as

$$n_e = \frac{\text{interconnected void volume}}{\text{total rock volume}}$$

$$v = \frac{K\frac{\Delta h}{\Delta l}}{n_e}$$

$v$ = average linear velocity of flow (cm/s)

**Figure 22.2**

Schematic representation of contaminant plumes possibly associated with various types of waste disposal.
*National Research Council, 1984, page 7,* Groundwater Contamination.

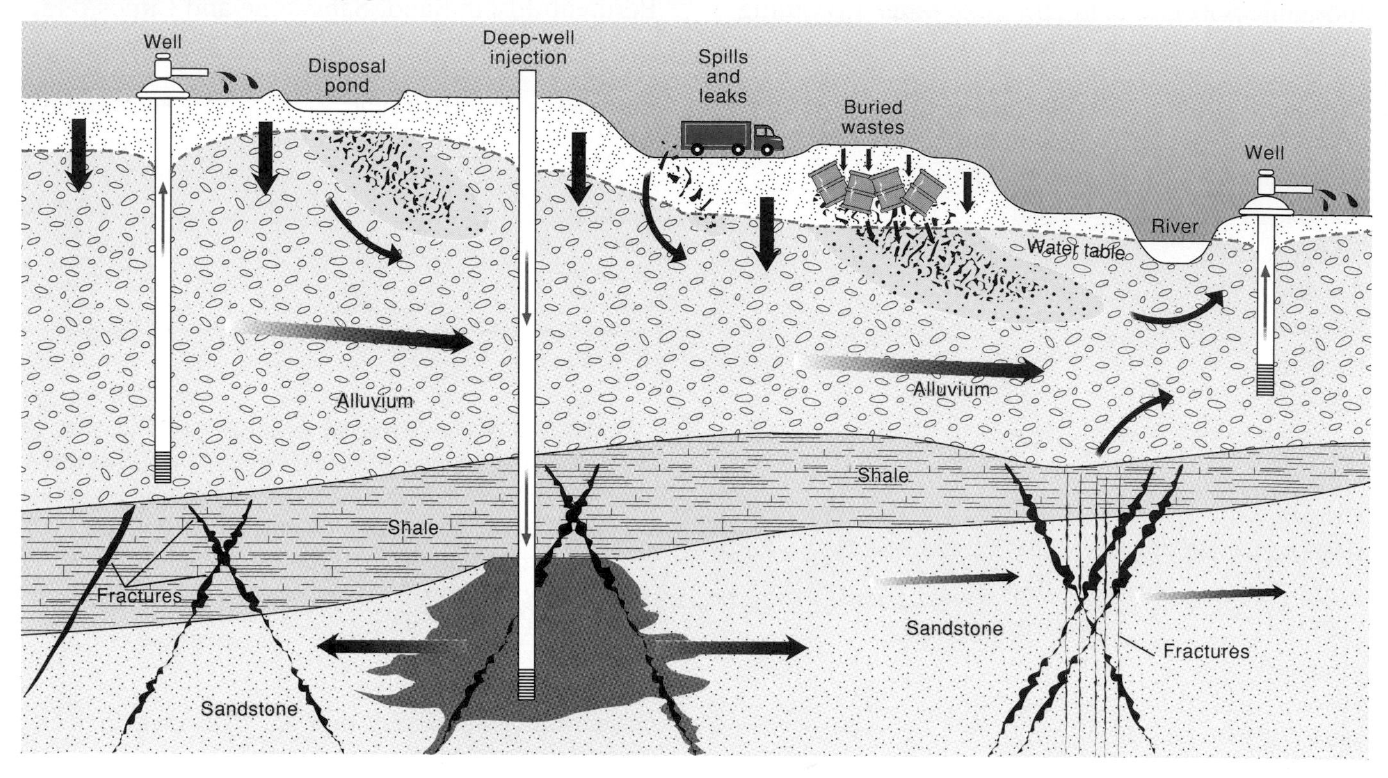


Most aquifers are not homogeneous, however, making times for contaminant migration through subsurface materials more difficult to calculate accurately, although we can often make reasonable estimates by using empirically obtained "retardation factors." Complications arise from variable geologic conditions (figure 22.3) and from the many types and characteristics of the contaminants themselves. Layered beds and lenses of less permeable rock within an aquifer can cause fingering and separation of a contaminant plume, and the clastic grains that form the aquifer cause the pollutant to disperse and be diluted (figure 22.4) within the aquifer. Pollutant-absorbing materials such as clay minerals might be irregularly distributed within the host rock. Pollutants from different sources might react chemically within an aquifer to create new substances, or microbes might interact with and reduce the amounts of some pollutants. Some liquid pollutants, such as brine, that are denser than water, sink and concentrate along the base of the aquifer rock; others that are less dense, such as gasoline, float and concentrate toward the top. Unfortunately, variation in both geologic conditions and the pollutants themselves occurs below the surface, where we cannot observe it directly. Groundwater becomes

**Figure 22.3**

Possible consequences of subsurface injection at a site not having the necessary hydrogeologic characteristics.
*Source: U.S. Geological Survey Water-Supply Paper 2281, 1986.*

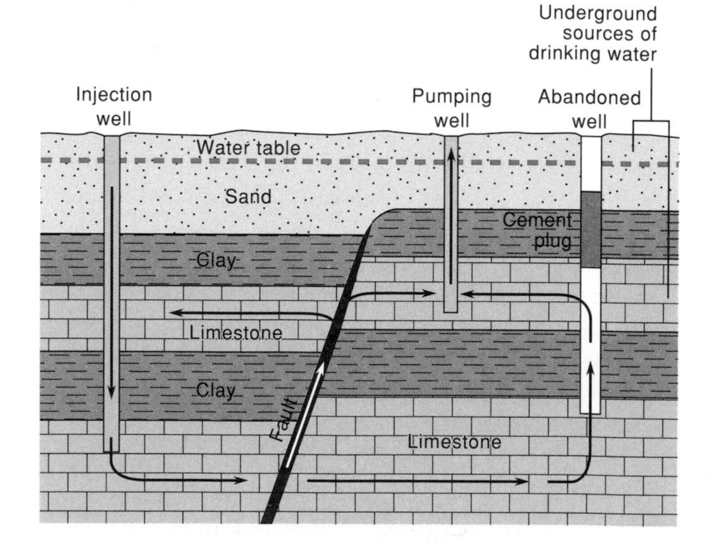

## Figure 22.4

(a) Relationship between the true flow path of a fluid in a clastic rock and the mean flow path. (b) Dispersion of the flow in the rock, illustrating the way an initially narrow band of pollutant spreads throughout the rock layer.

*R. Allen Freeze/John A. Cherry,* Groundwater, *Copyright © 1979, pp. 70, 384. Reprinted by permission of Prentice Hall, Englewood Cliffs, New Jersey.*

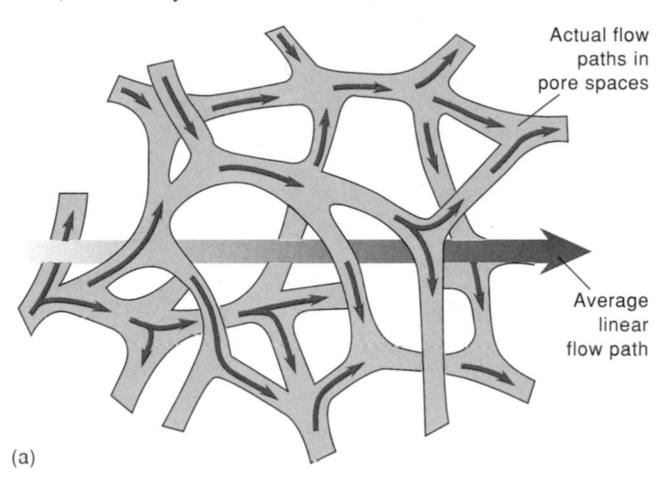


(a)

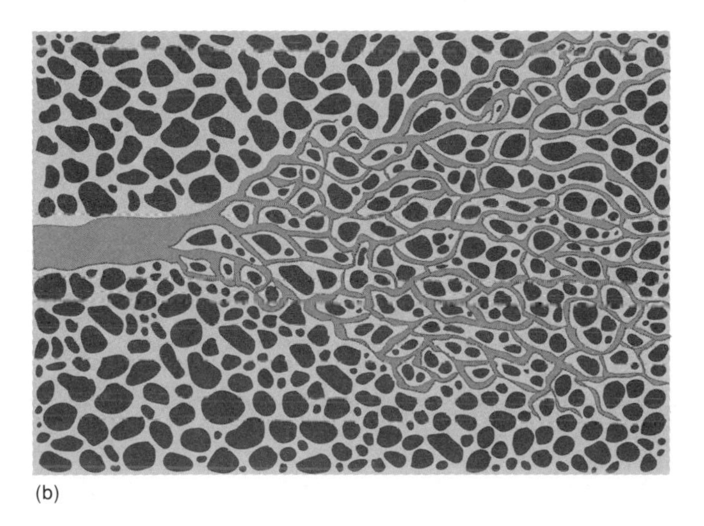

(b)

## Figure 22.5

Map of an area two miles to the east of Milleville showing surface drainage, elevation of the water table at 24 locations, and several man-made features.

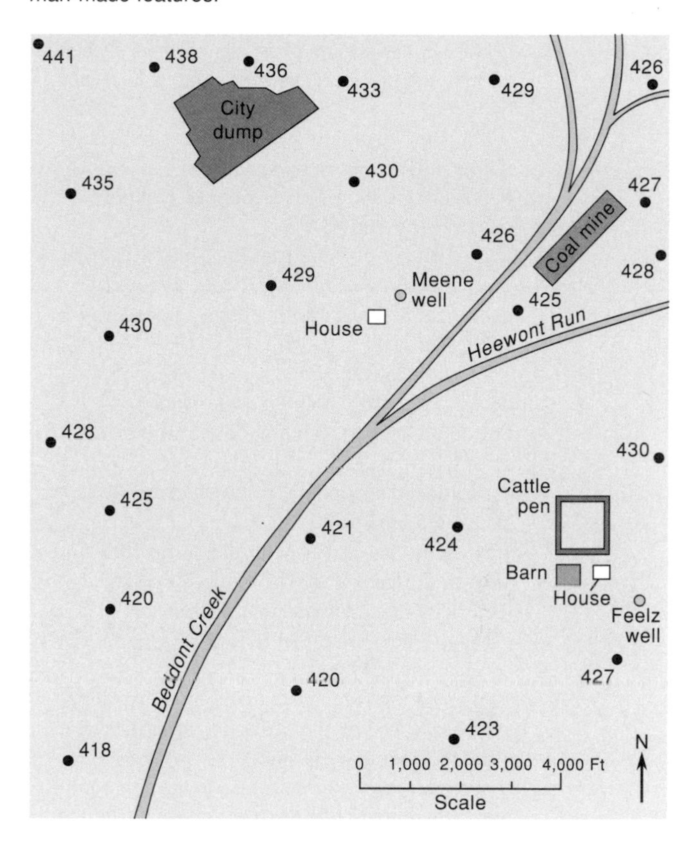


very difficult or even impossible to purify once it has been contaminated. And even if we have the technology to decontaminate an aquifer, the financial cost may well be so high that we must instead write off the aquifer as permanently unusable.

## Problems

1. What would be the advantages and disadvantages of locating a sanitary landfill on a river floodplain?
2. A well on your property draws water from the water table at an elevation of 600 m in a confined, homogeneous aquifer. At a distance of 9 km from your well, directly up the slope of the water table, is another well in which the water table is at 617 m. Contaminated wastewater has been dumped into this second well. The aquifer has a uniform effective porosity of 20% and a permeability of 100 millidarcies. Gravitational acceleration is 980 cm/s$^2$, the viscosity of water at 20°C is 0.01 gm/cm s, and water density is 1.00 gm/cm$^3$.
   a. How long will it take for the plume of contaminated water to reach your well?
   b. Would the presence of clay minerals in the aquifer be of interest to you? Why or why not?
   c. Suppose that two years from now some people approach you about purchasing your property. Their only question about water concerns the amount available from the well. What will you tell them about the contamination problem?
3. Figure 22.5 is a map of an area underlain by a moderately sorted, fine-grained sandstone that contains occasional lenses of clay. Based on the static water levels in wells not shown on the map, the locations of natural springs, the geologic map of the area, and scattered data from previous geologic investigations, the elevation of the water table is known at 24 locations

in the map area. It ranges from 418 ft in the southwest corner of the map to a maximum of 441 ft in the northwest corner. At most of these 24 sites, the water table is 10–12 ft below the ground surface. Based on these data,

a. Draw a contour map of the elevation of the water table in the map area, using a contour interval of two feet.
b. What is the hydraulic gradient between the city dump and Mr. Meene's house?
c. In which direction does Beddont Creek flow? How do you know?

Also located on the map are four man-made features:

—A cattle ranch owned by Mr. Feelz, a retired geologist, that has a pen, a barn, a house, and a water well.
—A small homestead owned by Mr. Meene, who draws water from both his water well and Beddont Creek, depending on the amount of seasonal rainfall.
—An abandoned strip mine, from which coal was excavated until 1980.
—A city dump that serves Millville, a small town just west of the map area. The dump is operated by a locally owned company and is excavated to a depth of about eight feet below the ground surface.

As a local environmental specialist in good standing, you have been asked by the Millville authorities to answer the following questions:

d. Is there any danger that refuse in the city dump will contaminate Mr. Meene's or Mr. Feelz's groundwater supply? Why or why not?
e. If you believe that either person's well might become contaminated from the dump, what would you recommend? Must the dump be abandoned? Could it still be used for certain types of refuse? Should a new dump be situated in the map area? If so, where would you recommend it be located and why?
f. Is Mr. Meene's water supply in danger of contamination from Mr. Feelz's cattle pen? Explain.
g. Might the abandoned coal mine contaminate either person's water supply? If so, which person is more likely to be harmed?

## Further Reading/References


Dee, N., McTernan, W. F., and Kaplan, E. (eds.), 1987, *Detection, Control, and Renovation of Contaminated Ground Water.* New York, American Society of Civil Engineers, 213 pp.

Hamilton, P. A., and Shedlock, R. J., 1992, *Are Fertilizers and Pesticides in the Groundwater?* U.S. Geological Survey Circular 1080, 15 pp.

LeBlanc, D. R., 1984, *Sewage Plume in a Sand and Gravel Aquifer, Cape Cod, Massachusetts.* U.S. Geological Survey Water-Supply Paper 2218, 28 pp.

Lloyd, O. B., Jr., and Reid, M. S., 1990, *Evaluation of Liquid Waste-Storage Potential Based on Porosity Distribution in the Paleozoic Rocks in Central and Southern Parts of the Appalachian Basin.* U.S. Geological Survey Professional Paper 1468, 81 pp.

Molenaar, D., 1988, *The Spokane Aquifer, Washington: Its Geologic Origin and Water-bearing and Water-quality Characteristics.* U.S. Geological Survey Water-Supply Paper 2265, 74 pp.

Moore, J. W., and Ramamoorthy, S., 1984, *Heavy Metals in Natural Waters.* New York, Springer-Verlag, 268 pp.

National Research Council, 1984, *Groundwater Contamination.* Washington, D.C., National Academy Press, 179 pp.

Patrick, R., Ford, E., and Quarles, J., 1987, *Groundwater Contamination in the United States,* 2nd ed. Philadelphia, University of Pennsylvania Press, 513 pp.

Sawhney, B. L., and Brown, K. (eds.), 1989, *Reactions and Movement of Organic Chemicals in Soils.* Madison, Wisconsin, Soil Science Society of America Special Publication No. 22, 474 pp.

U.S. Environmental Protection Agency, 1991, *Fact Sheet, Drinking Water Regulations Under the Safe Drinking Water Act,* June, 1991, Washington, D.C.

Yanggen, D. A., and Webendorfer, B., 1991, *Groundwater Protection through Local Land-Use Controls.* Wisconsin Geological and Natural History Survey Special Report 11, 48 pp.

exercise

# TWENTY-THREE

# Acid Rain

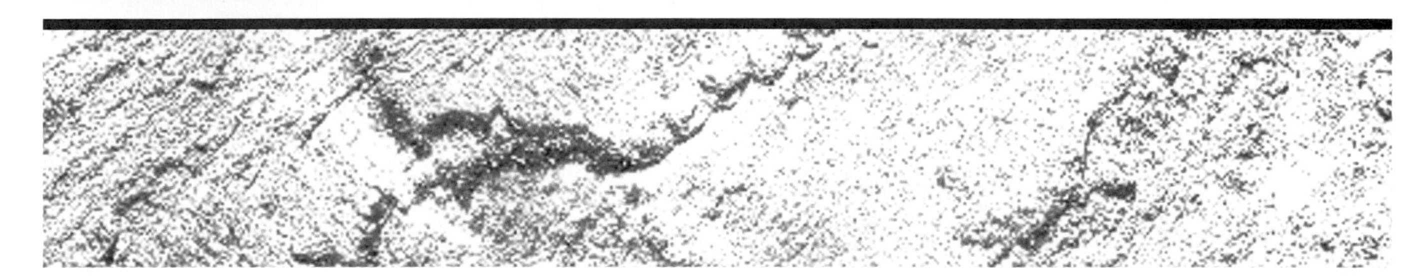

In recent years the term "acid rain" has become widely used in both the scientific literature and the popular press. Mention of the term conjures up images of fishless lakes, dying forests, and other biological disasters. But acid rain also has geologic effects on the bedrock, sediment, and soil that underlie and surround the affected lakes, streams, and aquifers. As is clear from our knowledge of weathering reactions, rainwater flows through the materials of the lithosphere, reacts with them, and causes chemical changes in both the waters and the solids.

## *Origins of Acid Rain*

"Normal" rain is a liquid that contains not only $H_2O$ molecules but also other substances that dissolve in the raindrops as they fall through the atmosphere. In unpolluted areas the most important of these substances is carbon dioxide, an atmospheric gas that reacts with water to generate carbonic acid:

$$CO_2 + H_2O \rightarrow \underset{\text{carbonic acid}}{H_2CO_3} \rightarrow H^+ + \underset{\text{bicarbonate}}{HCO_3^-}$$

Rainwater thereby becomes a dilute solution of carbonic acid, with a pH of 5.6 rather than the neutral value of 7 (pH = –log $H^+$ concentration). Thus, *all* natural rainwater is acid rain, even without the addition of artificial pollutants. But rain's natural pH of 5.6 has been constant for many millions of years; plant life, animal life, and weathering processes of rocks have all adjusted to it.

The acid rain produced by artificial pollutants, however, has pH values significantly lower than 5.6; values below 2 have been measured in some areas, and pH values lower than 5 are known to be harmful to both plants and animals. Extremely low pH values result primarily from emissions from factories and power plants that burn fossil fuels and secondarily from automobiles. The harmful emissions are sulfur dioxide ($SO_2$) gas and gaseous oxides of nitrogen (NO, $N_2O$), all of which combine with atmospheric oxygen:

$$2SO_2 + O_2 + 2H_2O \rightarrow 4H^+ + 2SO_4^{-2}$$
$$2NO + O_2 + 2H_2O \rightarrow 4H^+ + 2NO_3^-$$
$$N_2O + O_2 + 3H_2O \rightarrow 6H^+ + 2NO_3^-$$
$$N_2O + 2O_2 + H_2O \rightarrow 2H^+ + 2NO_3^-$$

Hydrogen ions create acidity, and atmospheric oxidation of industrial sulfurous and nitrogenous gases significantly increases the hydrogen-ion content of rainwater in an area. In North America these gaseous emissions are concentrated in, but not limited to, the so-called "rust belt" of northeastern United States, centered in Michigan, Indiana, Ohio, and Pennsylvania.

The actual distribution of acid rain is complicated by atmospheric movement. The dominant winds in the eastern half of the United States blow toward the north and northeast (figure 23.1), creating serious acid-rain problems in eastern Canada, New York State, and New England. In the most seriously affected areas of Pennsylvania and New York, rainwater pH has decreased from an assumed normal average of 5.6 before 1900 to about 4.5 in 1955 and 4.1 in 1982 (figure 23.2).

We could completely solve the acid-rain problem by taking either of two steps: (1) removing the sulfurous and nitrogenous gases before they leave the smokestacks of the factories and power plants, or (2) closing down the factories and power plants. Unfortunately, either solution would have serious social, economic, and political repercussions. Many operative factories are old, and removing the harmful compounds before they are vented might not be economically feasible. Shutting down the factories would throw large numbers of people out of work, as well as virtually destroy an important segment of American industry. On the other hand, both people and the natural environment are being increasingly harmed by the smokestack emissions (figure 23.3). Few politicians want to be held responsible for the social disruption resolving the acid-rain problem will entail.

## *Geologic Effects of Acid Rain*

Acid precipitation, whether natural or induced, interacts with minerals in common reactions such as the following:

$$\underset{\text{calcite}}{CaCO_3} + H^+ \rightarrow \underset{\text{bicarbonate}}{HCO_3^-} + Ca^{+2}$$

$$\underset{\text{calcic plagioclase feldspar}}{2CaAlSi_2O_8} + 4H^+ + 7H_2O \rightarrow \underset{\text{kaolinite clay}}{Al_2Si_2O_5(OH)_4} +$$

$$\underset{\text{soluble silica}}{2H_4SiO_4} + \underset{\text{soluble alumina}}{2Al(OH)_3^0} + 2Ca^{+2}$$

$$\underset{\text{olivine}}{MgSiO_4} + 2H^+ + 2H_2O \rightarrow \underset{\text{soluble silica}}{H_4SiO_4} + Mg^{+2} + 2OH^-$$

$$OH^- + H^+ \rightarrow H_2O$$

Hydrogen ions are consumed in such chemical reactions; they enter the reaction but do not leave. Hydrogen atoms exit the reactions as part of more complex atomic arrangements and can no longer cause strong acidity. Acidity is thus neutralized, making surface and underground waters less acid than their parent rainwater.

Chemical reactions with silica-poor minerals such as calcite, calcic plagioclase, or olivine can consume large amounts of hydrogen ions. Calcite is particularly effective because it is abundant (limestone), very soluble, and reacts much more rapidly than do silicate minerals. Calcic plagioclase and olivine are among the more common, fast-reacting silicate minerals; they also neutralize two moles of hydrogen ions for each mole of the mineral consumed in the reaction. Limestones, dolostones, and mafic igneous and metamorphic rocks contain large amounts of fast-reacting, hydrogen-consuming, silica-poor minerals.

**Figure 23.1**

Annual prevailing wind directions in the eastern United States.
*From W. D. Bischoff, et al.,* Geological Aspects of Acid Rain Deposition, *edited by O. P. Bricker. Copyright © 1984 Butterworth Publishers, Boston, MA. Reprinted by permission.*

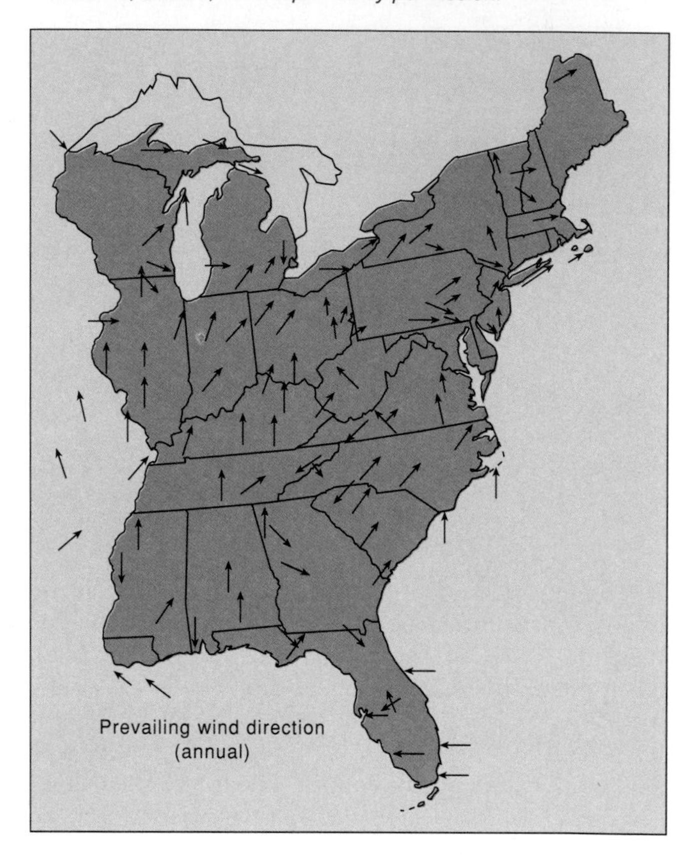


Silica-rich minerals are much slower to react in acid solutions, and they consume fewer hydrogen ions when they do react, as shown by the reactions of potassic and sodic feldspars:

$$2(K,Na)AlSi_3O_8 + 2H^+ + 9H_2O \rightarrow Al_2Si_2O_5(OH)_4 + 4H_4SiO_4 + 2(K,Na)$$

In these reactions, only one hydrogen ion is neutralized for each mole of feldspar consumed. Quartz, the most silica-rich mineral, neutralizes no hydrogen ions when it reacts with water:

$$SiO_2 + 2H_2O \rightarrow H_4SiO_4$$

Figure 23.2

Annual mean value of pH in precipitation, weighted by the amount of precipitation in the United States and Canada.
*Source: Environmental Protection Agency.*

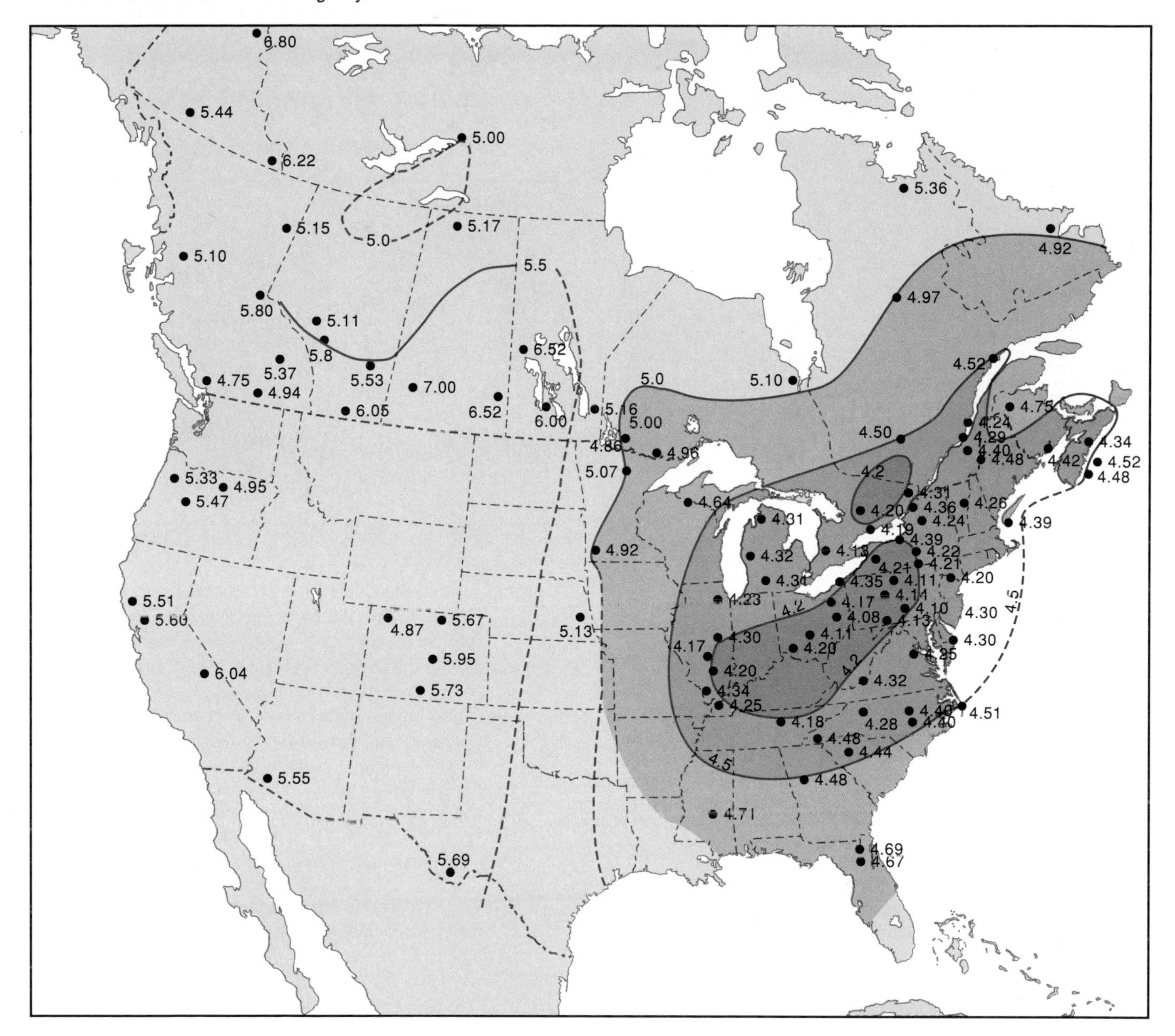


Hence, silica-rich rocks such as granite, quartz sandstone, quartzite, and most gneisses have relatively little ability to neutralize acid waters. Highly siliceous rocks are widespread throughout the Precambrian rocks of Scandinavia, Canada, New England, the Appalachians, and the Adirondack Mountains in New York State (figure 23.4). In these areas freshwater lakes located on the silica-rich rocks commonly have pH values of between five and three. Fish populations in such lakes have been decimated or severely reduced, as have trees in the surrounding forests.

Acid rain has adverse effects not only on lakes and natural vegetation, but also on soils and agriculture. Blueberries prefer acid soils, but most crops do not, so farmers must neutralize very acidic soils by adding lime (powdered limestone, $CaCO_3$). This process is costly for large farms. Furthermore, because rainwater flows after it reaches the ground, lime must be added to the soil regionally rather than locally.

**Figure 23.3**

High-altitude aerial photo of parts of Germany and Czechoslovakia showing the effects of acid rain (in false colors).

**Figure 23.4**

Relative sensitivities to acid rain in the conterminous United States. More sensitive areas contain higher proportions of silica-rich bedrock. Less sensitive areas tend to be underlain by limestone or dolostone, which have very high capacities for neutralizing acid rain. Whether acid rain is a serious problem depends not only on bedrock sensitivity, but also on the location of sulfurous and nitrogenous gas sources, atmospheric circulation patterns, and rainfall patterns.

*Source: Environmental Protection Agency.*

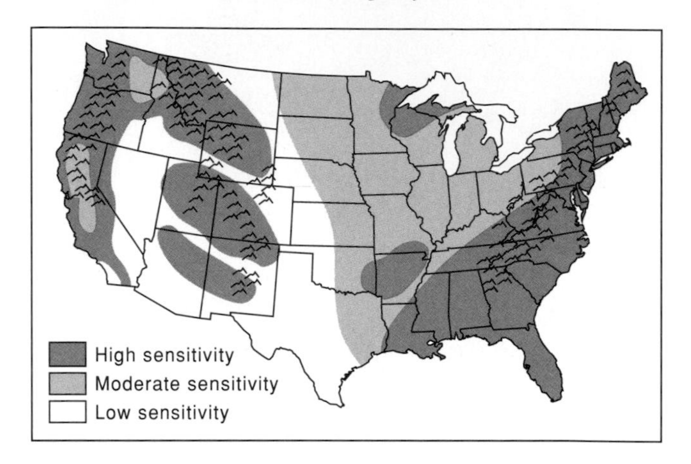


Because rainwater penetrates both the soil and the underlying porous, permeable bedrock, acid rain can seriously affect drinking-water supplies. As is true of reactions at the ground surface, limestone aquifers quickly neutralize excess acidity in groundwaters, but quartzose sandstones do not. Acid rain often liberates heavy metals normally held immobile on clay minerals in the soil; the metals then can move downward into groundwater used for human consumption, causing harmful effects.

## Problems

1. Litmus, a coloring agent made from certain lichens, is the essential component of litmus paper, which is used to determine the pH of liquids. Litmus paper turns red in acidic solutions and blue in basic ones. Modern refinements of litmus paper commercially available for under $20 per package provide a sharp color change for each 0.5 pH unit from pH 0 to pH 14.

   Using the litmus paper provided, determine the pH of each of the following:
   a. tap water
   b. bottled water
   c. vinegar
   d. tap water as a student blows into it with a straw
   e. water containing powdered gypsum
   f. water plus baking soda

   Explain why the pH is acidic or basic for each measurement.
2. From a base pH of 5.6 at the turn of the century, the rainwater pH in parts of the northeastern United States has decreased to 4.1. What percentage increase is this in $H^+$ concentration?
3. What would be the effect on acid-rain distribution of increasing smokestack heights at factories that emit sulfurous gases?
4. How is reactivity to acid rain related to mineral composition of igneous rocks?
5. Collect small bottles of water from local lakes, streams, and rainwater in your neighborhood and determine their pH values. Explain your results.
6. Explain briefly the relationship between local geology and problems caused by acid rain.

## Further Reading/References


Johnson, N. M., 1984, Acid rain neutralization by geologic materials: in *Geologic Aspects of Acid Deposition,* O. P. Bricker, (ed.). Boston, Butterworth, pp. 37–54.

Kahan, A. M., 1986, *Acid Rain: Reign of Controversy.* Golden, Colorado, Fulcrum, 238 pp.

Likens, G. E., Wright, R. F., Galloway, J. N., and Butler, T. J., 1979, Acid rain: *Scientific American,* v. 241, no. 4, pp. 43–51.

Park, C. C., 1987, *Acid Rain: Rhetoric and Reality.* New York, Methuen, 272 pp.

Regens, J. L., and Rycroft, R. W., 1988, *The Acid Rain Controversy.* Pittsburgh, University of Pittsburgh Press, 228 pp.

Turk, J. T., 1983, *An Evaluation of Trends in the Acidity of Precipitation and the Related Acidification of Surface Water in North America.* U.S. Geological Survey Water-Supply Paper 2249, 18 pp.

Wellburn, A., 1988, *Air Pollution and Acid Rain: The Biological Impact.* New York, John Wiley & Sons, 274 pp.

exercise

# TWENTY-FOUR

## *Radon in the Environment*

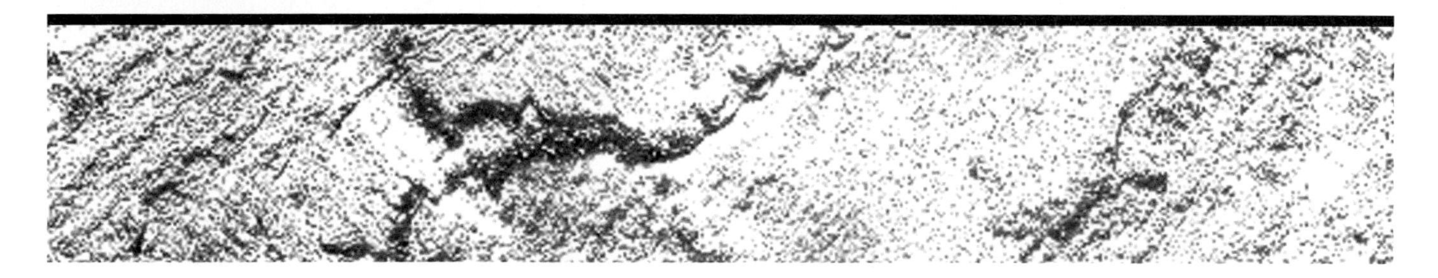

For some types of environmental dangers, both cause and effect are obvious. The clearest example is an oil spill, where we can see a fractured ship, black oil on the sea surface, and damaged marine organisms. Less visible are organic pollutants in drinking water, colorless compounds for which presence, sources, and effects may be difficult to detect. Among the least obvious dangers is radon, a naturally occurring, colorless, odorless, and tasteless radioactive gas found in some homes. There is no evidence linking such radon gas exposure to cancer; however, the known links between other radioactive substances and cancer has raised concerns about the possible effect of long-term radon exposure. Though the extent of the indoor radon problem is not yet fully documented, surveys indicate that 6%—6 million—U.S. homes might have high levels of radon.

Radon is the only gas among the 13 radioactive elements produced as intermediaries during the decay of uranium-238 to lead-206 (table 24.1). Uranium and its decay products are not distributed uniformly among the different types of rocks. Among igneous and metamorphic rocks, silica-rich varieties such as granite and quartzo-feldspathic schists and gneisses contain four times more uranium than do basalts or gabbros (table 24.2). This uranium resides largely in the mineral zircon. As the rocks weather, the zircon grains are released. Chemical alteration of other uranium-bearing minerals also occurs, mobilizing any uranium in contact with atmospheric oxygen. Rivers or groundwater transport the uranium until they move into an oxygen-deficient environment, where the uranium becomes insoluble and precipitates, usually as an ion adsorbed on organic matter. Thus, sediments rich in organic matter often become enriched in uranium—and, therefore, in radon. Soils formed on uranium-rich rocks may become similarly enriched in radioactive elements, although in permeable, oxygenated soils the uranium can be remobilized and carried away (perhaps toward someone else's house?). In summary, radon concentrations are relatively high in granites, in quartzo-feldspathic schists and gneisses, in sandstones derived from granites (arkoses), and in organic-rich sedimentary materials such as black shales and phosphorites (rocks rich in chemically precipitated calcium phosphate). The federal government and several states have published maps of radon-potential based on a combination of actual measurements and outcrop patterns of rock types.

TABLE 24.1

Decay Path of $^{238}U$

| Isotope | Half-life* | Principal Decay Modes |
|---|---|---|
| $^{238}U$ | $4.5 \times 10^9$ yr | α,γ |
| $^{234}Th$ | 24.1 day | β,γ |
| $^{234}Pa$ | 6.75 hr | β,γ |
| $^{234}U$ | $2.48 \times 10^5$ yr | α,γ |
| $^{230}Th$ | $8.0 \times 10^4$ yr | α,γ |
| $^{226}Ra$ | 1, 622 yr | α,γ |
| $^{222}Rn$ | 3.82 day | α |
| $^{218}Po$ | 3.05 min | α |
| $^{214}Pb$ | 26.8 min | β |
| $^{214}Bi$ | 19.7 min | α,β,γ |
| $^{214}Po$ | $1.6 \times 10^{-4}$ sec | α |
| $^{210}Pb$ | 22.0 yr | β,γ |
| $^{210}Bi$ | 5.01 day | β |
| $^{210}Po$ | 138.4 day | α |
| $^{206}Pb$ | stable | α |

*The time required for half the atoms of a radioactive substance to decay. Alpha, beta, and gamma are the three types of emissions given off during the decay process.

TABLE 24.2

Averages and Ranges of Uranium Contents (ppm) for Common Rock Types

| Rock Type | Mean | Range |
|---|---|---|
| Igneous | | |
| Gabbro, basalt | 0.8 | 0.1–3.5 |
| Diorite, quartz diorite | 2.5 | 0.5–12 |
| Granite, rhyolite | 4.0 | 1.0–22 |
| Sedimentary | | |
| Orthoquartzite | 0.5 | 0.2–0.6 |
| Sandstone | 1.5 | 0.5–4 |
| Carbonate | 1.6 | 0.1–10 |
| Shale | 3.0 | 1.0–15 |
| Black shale | | 3.0–1,250 |
| Lignite | | 10.0–2,500 |
| Phosphorite | | 50.0–2,500 |

Uranium occurs in minor or trace amounts in all rocks and soils, and radon generated in the top 10–20 ft of the ground either decays to a solid ($^{218}Po$) in the ground or escapes to the air. In the air, the radon generally dilutes to very low concentrations before decaying. Average levels of radon emitted from rock and soil are about 0.2 picocuries per square meter of soil (table 24.3). (One curie [ci] is defined as $3.7 \times 10^{10}$ isotopic disintegrations per second; a picocurie [pCi] is $10^{-12}$ curie.) Radon can accumulate in buildings, however, constituting a health hazard if the level becomes too high. But what is too high? The Environmental Protection Agency currently recommends remedial action for levels above 4 pCi per liter of air, a level of exposure equivalent to 200 X-rays per year in a physician's office. However, this EPA guideline might be set much too high; physicians suggest limiting patient X-rays to one or two per year, if possible. This more conservative limit would suggest a maximum radon level of only 0.04 pCi, a goal unattainable in most homes. Radon might simply be one of the unavoidable environmental factors that limits the average human lifespan to under 100 years.

Radon enters homes through three major pathways (figure 24.1): radon-containing groundwater pumped into homes through wells; radon gas that migrates upward from soil and rock and enters houses through porous or cracked cement, often in basements; and construction materials such as building blocks made of substances that emit radon gas. Brick, stone, concrete, and other building materials that contain trace amounts of uranium can release radon gas. Radon gas builds to more dangerous levels in well-insulated houses because it cannot escape. Insulation might conserve the nation's petroleum, natural gas, and coal supplies, but without adequate ventilation, it might also be bad for your longevity. In general, radon concentrations are highest in basements and lower floors and decrease higher above ground level. The radon particles attach themselves to dust, which people then inhale. The particles then adhere to the surface of the lungs, releasing radiation to the surrounding tissues, and the radon decays to other radioactive daughter products.

## TABLE 24.3

### Radon Risk-Evaluation Chart

| Annual Radon level: | If a community of 100 people were exposed to this level: | This risk of dying from lung cancer compares to: |
|---|---|---|
| 100 pCi/l | About 35 people in the community may die from radon | Having 2,000 chest X-rays each year |
| 40 pCi/l | About 17 people in the community may die from radon | Smoking 2 packs of cigarettes each day |
| 20 pCi/l | About 9 people in the community may die from radon | Smoking 1 pack of cigarettes each day |
| 10 pCi/l | About 5 people in the community may die from radon | Having 500 chest X-rays each year |
| 4 pCi/l | About 2 people in the community may die from radon | Smoking 1/2 pack of cigarettes each day |
| 2 pCi/l | About 1 person in the community may die from radon | Having 100 chest X-rays each year |

Levels as high as 3,500 pCi/l have been found in some homes. The average radon level outdoors is around 0.2 pCi/l or less.
The risks shown in this chart are for the general population, including men and women of all ages as well as smokers and nonsmokers. Children may be at a higher risk.
*Source: Environmental Protection Agency.*

## Problems

1. The major source of domestic drinking water near Carson City, the capital of Nevada, is shallow groundwater from neighboring Carson Valley. Consequently, the U.S. Geological Survey surveyed radon contents in the groundwater at the request of the state government. The areal geography, sampling sites, and survey results are shown in figure 24.2A. A cross-section of the local topography and valley fill are shown in figure 24.2B, and the generalized geology of the area in figure 24.2C. Use this information to answer the following questions:
   a. The drainage basin for the water ("hydrographic area") is larger east of the Carson River than to the west. Would you have expected this result, based on the appearance of figure 24.2A? Why or why not?
   b. How many potential aquifers does the cross-section show?
   c. Which of these potential aquifers do you think is the major water source for private homes? Why?
   d. What part of Carson Valley has the highest radon levels?
   e. Explain this result in terms of the geology of the area.

**Figure 24.1**

Major radon entry routes into houses.
*Source: Environmental Protection Agency.*

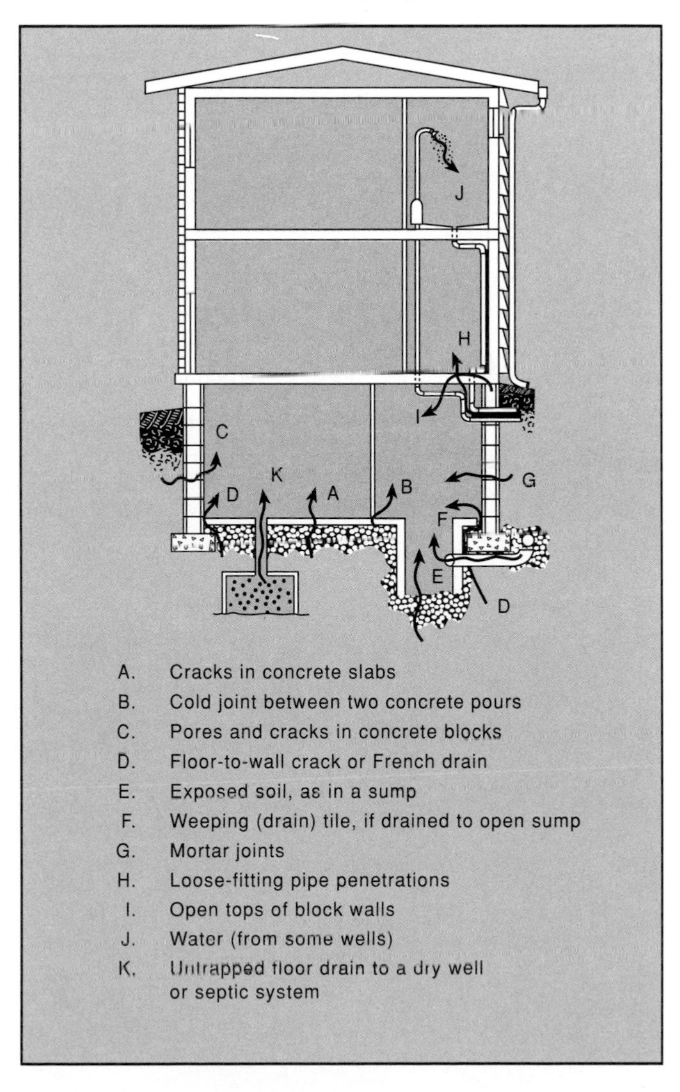

## Figure 24.2a

Location of study areas and sampling sites. Symbols indicate levels of radon-222 concentrations.
*Source: U.S. Geological Society.*

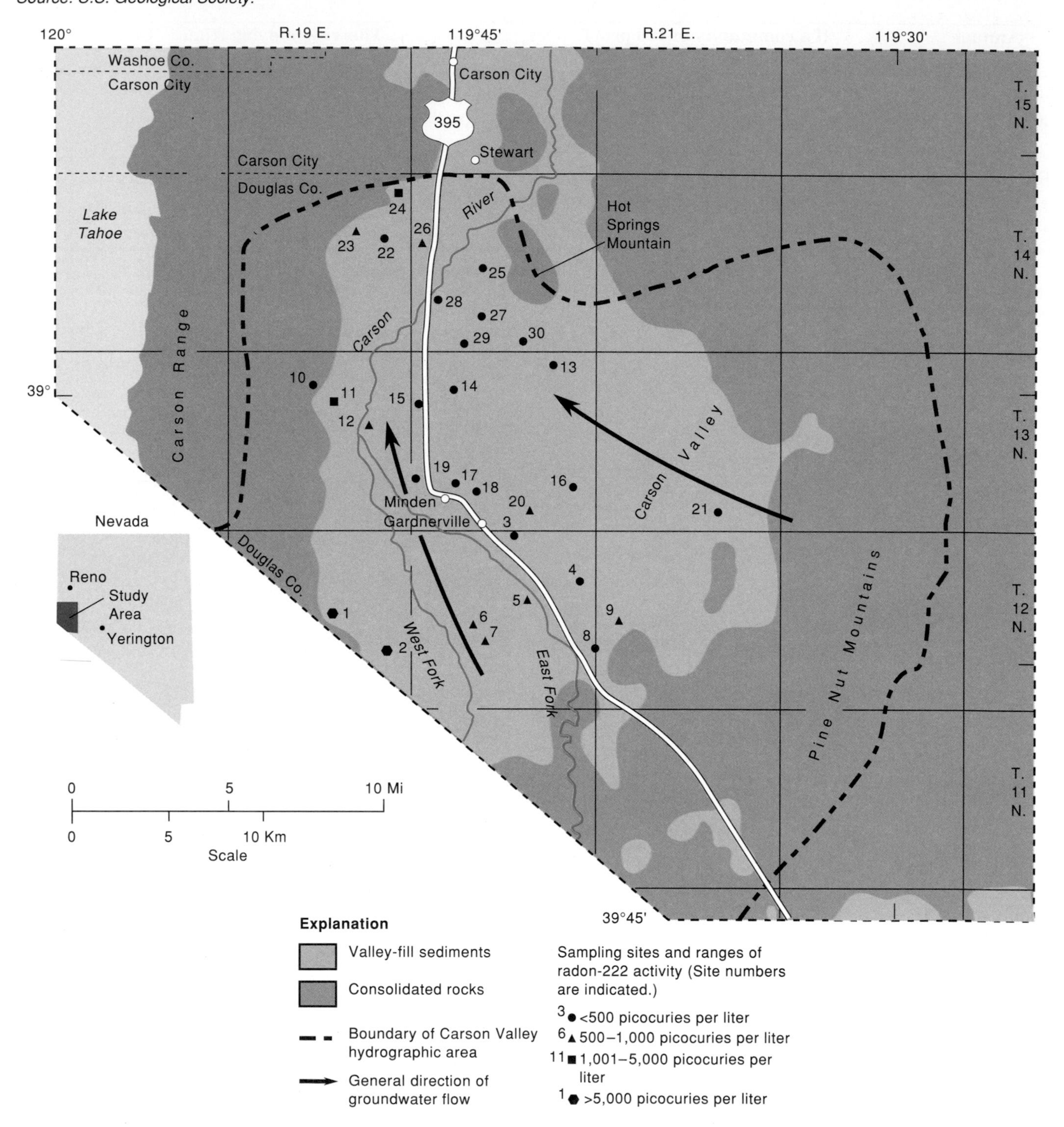

## Figure 24.2b

Cross-section of hydrogeologic relations in Carson Valley. Upper part of basin-fill section is Quaternary age and lower part is Tertiary age. View looking north.

*Source: U.S. Geological Society.*

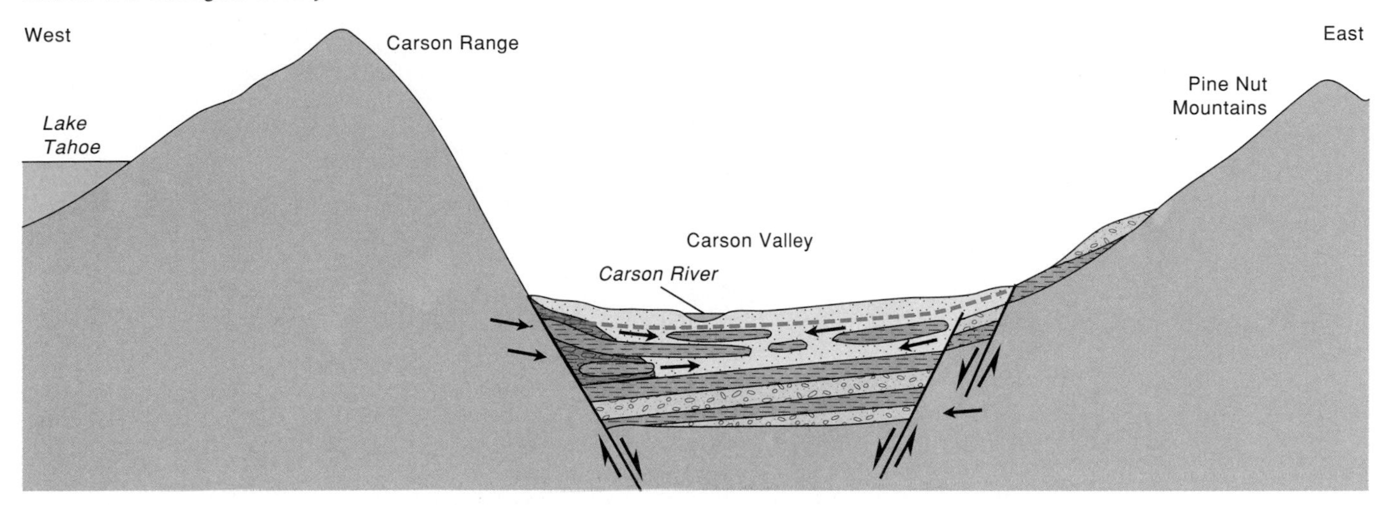


Explanation

Fluviatile and lacustrine deposits (groundwater can be confined or unconfined)

Dominated by cobbles and gravel

Dominated by sand and silt

Dominated by clay (thick units overlying bedrock are of Tertiary age)

Alluvial-fan deposits (poorly sorted: groundwater generally confined)

Bedrock

Major fault (arrows show relative direction of movement)

Groundwater flow path

Water table

2. Another recent study of radon occurrence involved rocks of Late Silurian, Early Devonian, and Middle Devonian age in Onondaga County of western New York State (figures 24.3, 24.4). Radon levels were measured in basements of 210 single-family houses and defined a radon belt that averaged 8.8 pCi/liter and paralleled the geologic strike in the area. The belts of rocks on both sides of this "hot" zone averaged only 1.2 and 2.5 pCi/liter. As the environmental geologist assigned to investigate and explain the radon levels shown on the map and cross-section, you must answer the following questions:
   a. In figure 24.3, the rock unit called "siltstone, sandstone, shale" is the youngest unit on the map and is also the most erosion-resistant. Sketch a cross-section along the line A-A′ based on this information.
   b. What does the use of a logarithmic scale for the X-axis do to the appearance of the data?
   c. If you were using radon occurrence as the sole criterion for situating your new house, on which formation would you build it?
   d. Suppose a location with low radon occurrence meant you would have to drive 20 miles each way to work. What would you do then? How about 50 miles each way? Suppose building the house at the low-radon location would cost 50% more than building it where the radon is more than 10 pCi/liter. What would you do then?
3. List the minerals and other substances in sedimentary rocks that are likely to contain and emit radon.
4. In terms of radiation dosages from radon, how many times more dangerous to your health is smoking a pack of cigarettes each day than having a chest X-ray once each year?
5. In what way does the permeability of a sedimentary rock affect its radon content?

## Figure 24.2c

Generalized geology of the study area and adjacent areas.
*Source: U.S. Geological Society.*

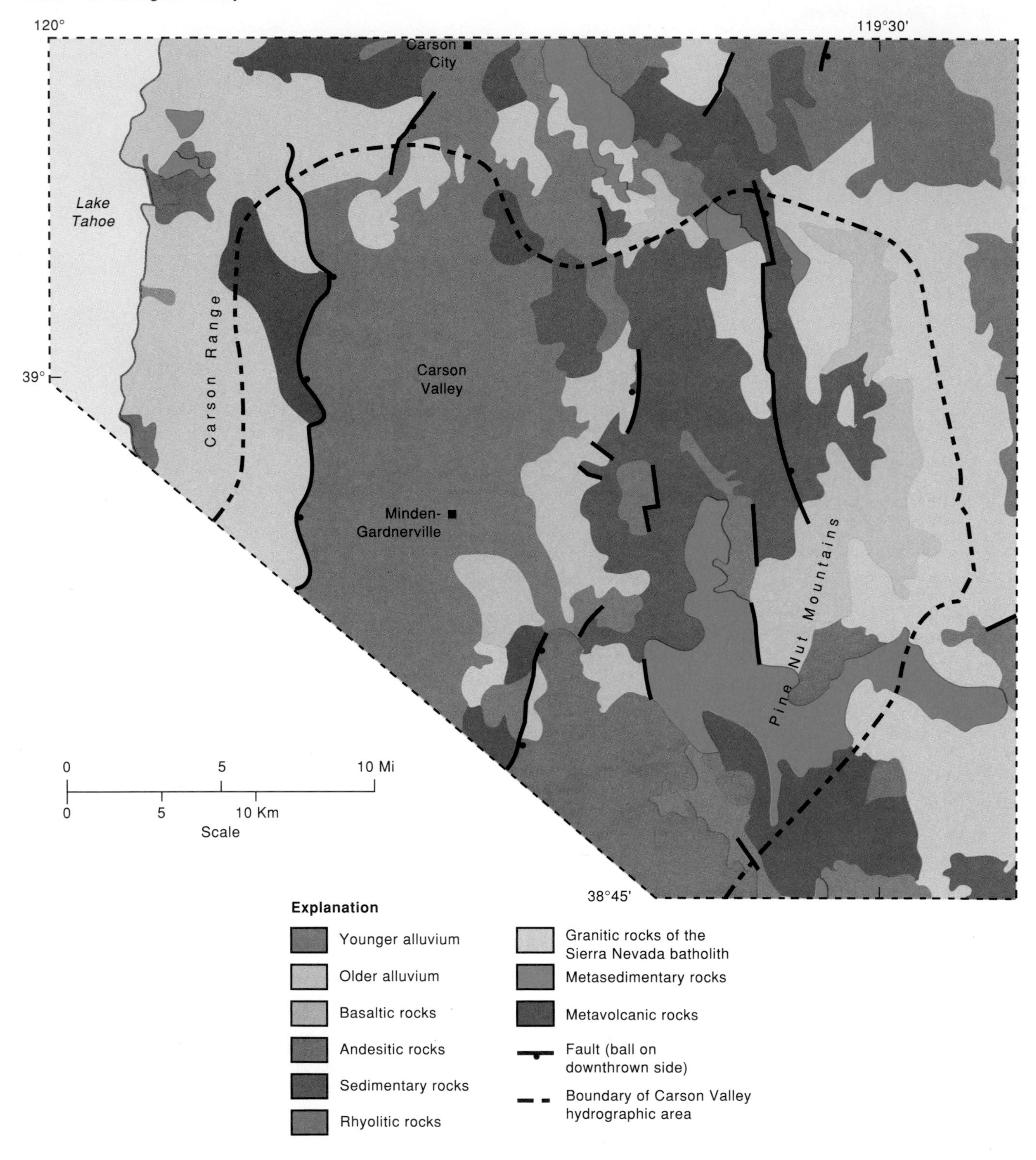

## Figure 24.3

Radon activities (pCi/l) in Onondaga County, New York, plotted on bedrock geologic map. Radon was measured in basements of single-family homes during the 1987–1988 heating season by using charcoal canisters to obtain four-day integrated samples. Bold numbers are geometric mean radon activities for regions A, B, and C.

*From B. M. Hand and J. E. Banikowski,* Geology, *16:775, 1988. Copyright © 1988 Geological Society of America, Boulder, CO. Reprinted by permission of the authors.*

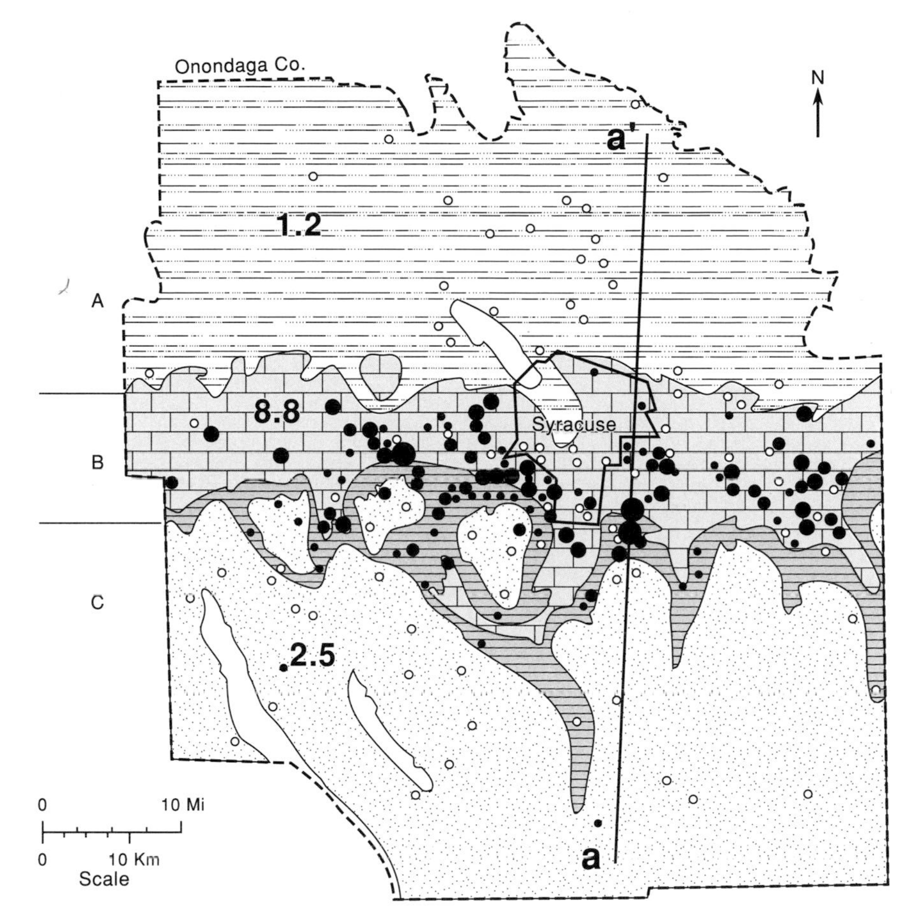


## Figure 24.4

Radon activities plotted against stratigraphic position. Vertical lines indicate geometric means; diagonal ruled pattern = area within 1 standard deviation of mean. A, B, C correspond to regions in Figure 24.3.

*From B. M. Hand and J. E. Banikowski,* Geology, *16:777, 1988. Copyright © 1988 Geological Society of America, Boulder, CO. Reprinted by permission of the authors.*

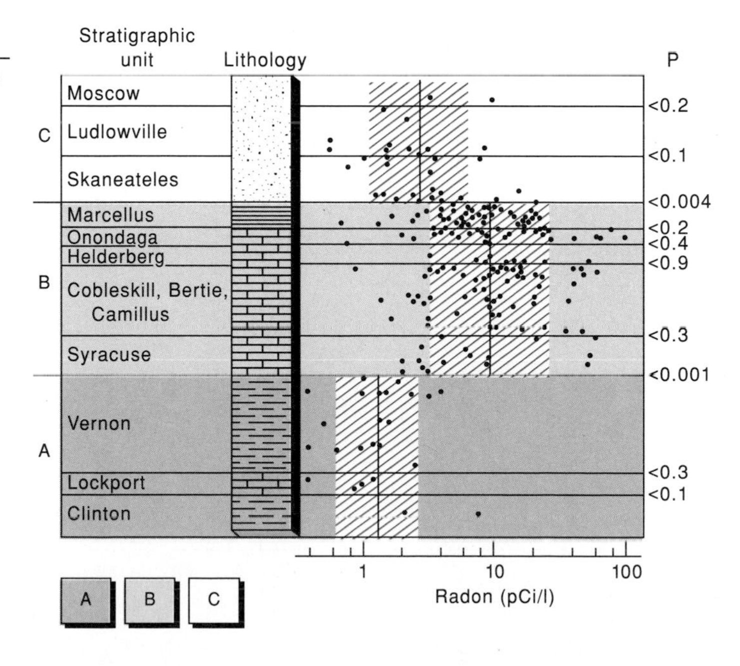

## Further Reading/References

Abelson, P. H., 1991, Mineral dusts and radon in uranium mines: *Science,* v. 254, No. 5033, p. 777.

Brookins, D. G., 1990, *The Indoor Radon Problem.* New York, Columbia University Press, 229 pp.

Cohen, B. L., 1987, *Radon: A Homeowner's Guide to Detection and Control.* Mt. Vernon, New York, Consumers Union, 215 pp.

Cothern, C. R., and Smith, J. E., Jr. (eds.), 1987, *Environmental Radon.* New York, Plenum, 363 pp.

Gundersen, L. C. S., and Wanty, R. B., 1991, *Field Studies of Radon in Rocks, Soils, and Water.* U.S. Geological Survey Bulletin 1971, 334 pp.

Hand, B. M., and Banikowski, J. E., 1988, Radon in Onondaga County, New York: paleohydrogeology and redistribution of uranium in Paleozoic sedimentary rocks: *Geology,* v. 16, pp. 775–778.

appendix

# A Mathematical Conversion Factors

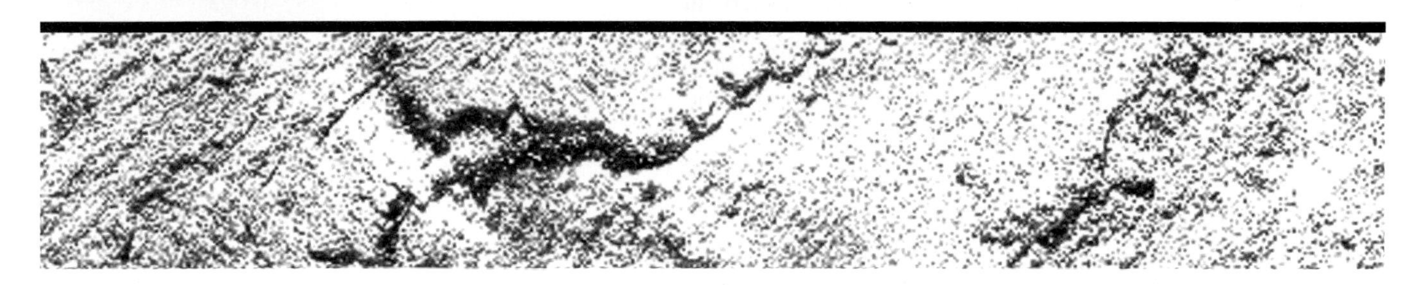

| To Convert from | To | Multiply by |
|---|---|---|
| Acre–feet | Gallons | $3.26 \times 10^5$ |
| Acre–foot | Cubic meters | 1,233.5 |
| Acres | Square feet | 43,560 |
| Barrels oil (bbl) | Cubic feet ($ft^3$) | 5.61 |
| Barrels oil | Gallons (gal) | 42 |
| Bars | Pounds/square inch ($lb/in^2$) | 14.504 |
| Centimeters (cm) | Inches (in) | 0.394 |
| Centimeters per second (cm/s) | Feet per day (ft/day) | 2,835 |
| Centimeters per second | Gallons per day per square foot ($gal/day/ft^2$) | 21,200 |
| Centimeters per second | Meters per day (m/day) | 864 |
| Cubic centimeters ($cm^3$) | Cubic inches ($in^3$) | 0.061 |
| Cubic inches ($in^3$) | Cubic centimeters ($cm^3$) | 16.387 |
| Cubic feet ($ft^3$) | Barrels of oil (bbl) | 0.18 |
| Cubic feet | Cubic meters ($m^3$) | 0.028 |
| Cubic feet per second ($ft^3/s$) | Cubic meters per second ($m^3/s$) | 0.003 |
| Cubic meters | Acre-feet | $8.11 \times 10^{-4}$ |
| Cubic meters ($m^3$) | Cubic feet ($ft^3$) | 35.249 |
| Cubic meters per second ($m^3/s$) | Cubic feet per second ($ft^3/s$) | 353.107 |
| Cubic miles ($mi^3$) | Cubic kilometers ($km^3$) | 4.167 |
| Cubic kilometers ($km^3$) | Cubic miles ($mi^3$) | 0.240 |
| Feet (ft) | Meters (m) | 0.305 |
| Feet per day (ft/day) | Centimeters per second (cm/s) | $3.53 \times 10^{-4}$ |
| Feet per mile (ft/mi) | Meters per kilometer (m/km) | 0.188 |
| Gallons | Acre-feet | $3.07 \times 10^{-6}$ |
| Gallons per day per square foot ($gal/day/ft^2$) | Centimeters per second (cm/s) | $4.72 \times 10^{-5}$ |
| Grams (g) | Ounces (oz) | 0.035 |
| Hectares (ha) | Square feet ($ft^2$) | $1.076 \times 10^5$ |
| Inches (in) | Centimeters (cm) | 2.540 |
| Kilograms (kg) | Pounds (lb) | 2.205 |
| Kilometers (km) | Miles (mi) | 0.621 |
| Liters (l) | Quarts (qt) | 1.057 |

| To Convert from | To | Multiply by |
| --- | --- | --- |
| Meters (m) | Feet (ft) | 3.281 |
| Meters | Yards (yd) | 1.094 |
| Meters per day (m/day) | Centimeters per second (cm/s) | 0.00116 |
| Meters per kilometer (m/km) | Feet per mile (ft/mi) | 5.283 |
| Miles (mi) | Kilometers (km) | 1.609 |
| Ounces (oz) | Grams (g) | 28.350 |
| Pounds (lb) | Kilograms (kg) | 0.454 |
| Quarts (qt) | Cubic centimeters ($cm^3$) | 946.358 |
| Quarts | Liters (l) | 0.946 |
| Square centimeters ($cm^2$) | Square inches ($in^2$) | 0.155 |
| Square feet ($ft^2$) | Square meters ($m^2$) | 0.093 |
| Square feet | Hectares (ha) | $0.929 \times 10^{-5}$ |
| Square inches ($in^2$) | Square centimeters ($cm^2$) | 6.452 |
| Square kilometers ($km^2$) | Square miles ($mi^2$) | 0.386 |
| Square meters ($m^2$) | Square yards ($yd^2$) | 1.196 |
| Square miles ($mi^2$) | Square kilometers ($km^2$) | 2.589 |
| Square yards ($yd^2$) | Square meters ($m^2$) | 0.836 |
| Yards (yd) | Meters (m) | 0.914 |
| | | |
| Meters (m) | Kilometers (km) | $10^{-3}$ |
| Meters | Centimeters (cm) | $10^{2}$ |
| Meters | Millimeters (mm) | $10^{3}$ |
| Meters | Micrometers (microns; µm) | $10^{6}$ |

appendix

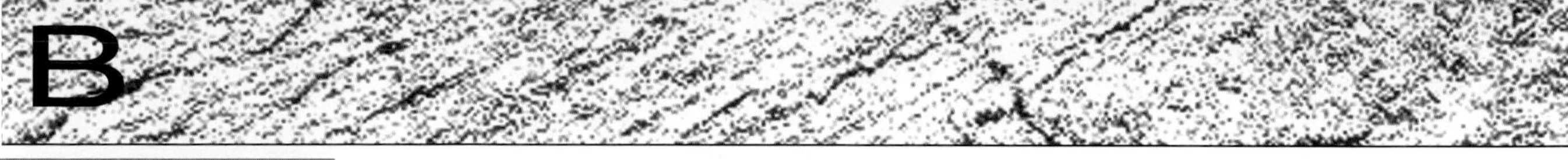

# B Geologic Time Scale

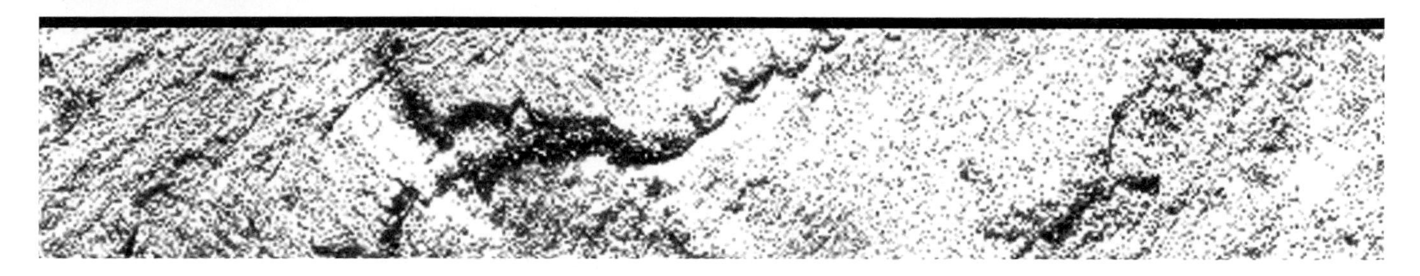

| Era | Period | Epoch | Start of Interval (Million Years Before Present) | Length of Interval | Percent of Geologic Time |
|---|---|---|---|---|---|
| | | Holocene | 0.01 | 0.01 | 0.0002 |
| | Quaternary | Pleistocene | 1.8 | 1.79 | 0.04 |
| | | Pliocene | 5 | 3.2 | 0.07 |
| Cenozoic | | Miocene | 23 | 18 | 0.39 |
| | Tertiary | Oligocene | 34 | 11 | 0.24 |
| | | Eocene | 53 | 19 | 0.41 |
| | | Paleocene | 65 | 12 | 0.26 |
| | Cretaceous | | 135 | 70 | 1.52 |
| Mesozoic | Jurassic | | 205 | 70 | 1.52 |
| | Triassic | | 250 | 45 | 0.98 |
| | Permian | | 290 | 40 | 0.87 |
| | Pennsylvanian | | 320 | 30 | 0.65 |
| | Mississippian | | 355 | 35 | 0.76 |
| Paleozoic | Devonian | | 410 | 55 | 1.20 |
| | Silurian | | 438 | 28 | 0.61 |
| | Ordovician | | 510 | 72 | 1.57 |
| | Cambrian | | 570 | 60 | 1.30 |
| Neoproterozoic | Cryogenian | | 850 | 280 | 6.09 |
| | Tonian | | 1,000 | 150 | 3.26 |
| | Stenian | | 1,200 | 200 | 4.35 |
| Mesoproterozoic | Ectasian | | 1,400 | 200 | 4.35 |
| | Calymmian | | 1,600 | 200 | 4.35 |
| | Statherian | | 1,800 | 200 | 4.35 |
| | Orosirian | | 2,050 | 250 | 5.43 |
| Paleoproterozoic | Rhyacian | | 2,300 | 250 | 5.43 |
| | Siderian | | 2,500 | 200 | 4.35 |
| Archean | | | 4,600 | 2,100 | 45.65 |
| | | | | 4,600 | 100.00 |

# appendix C

## Earth Science Information Centers

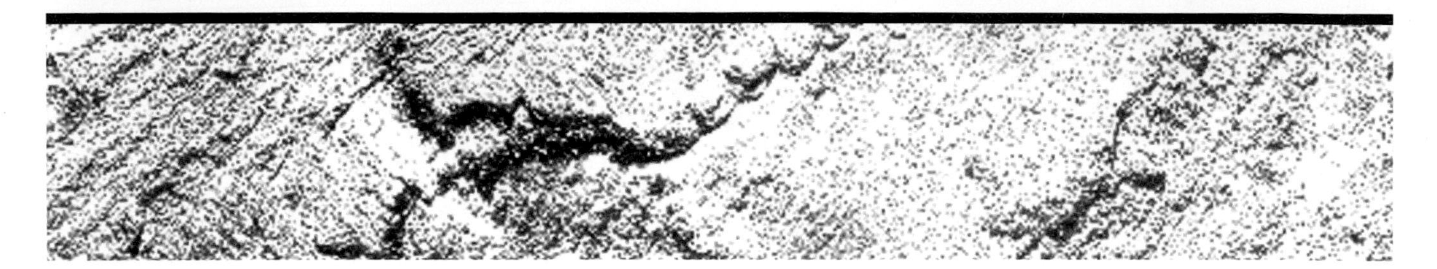

The Earth Science Information Centers (ESICs) offer nationwide information and sales service for United States Geological Survey (USGS) maps and earth-science publications. This ESIC network provides information about geologic, hydrologic, topographic, and land-use maps, books, and reports; aerial, satellite, and radar images and related products; earth science and map data in digital form and related applications software; and geodetic data. Write to any ESIC for addresses of the over 60 state ESIC offices.

Any ESIC can fill orders for custom products such as aerial photographs, orthophotoquads, digital cartographic data, and geographic names data.

ESICs can also provide information about earth-science materials from many public and private producers in the United States using automated catalog systems for information retrieval and research services.

For further information contact any ESIC or call 1–800–USA-MAPS.

**National Distribution Centers**

Lakewood–ESIC
Box 25046, Federal Center, MS 504
Building 25, Rm. 1813
Denver, CO 80225–0046
303–236–5829; FTS 776–5829

Reston–ESIC
507 National Center
Reston, VA 22092
703–648–6045; FTS 959–6045

Washington, D.C.–ESIC
U.S. Department of the Interior
1849 C Street, NW, Rm. 2650
Washington, D.C. 20240
202–208–4047; FTS 268–4047

**Regional Distribution Centers**

Anchorage–ESIC
4230 University Drive, Rm. 101
Anchorage, AK 99508–4664
907–786–7011; FTS 868–7011

Anchorage–ESIC
Room G-84
605 West 4th Avenue
Anchorage, AK 99501
907–271–2754; FTS 868–2754

Denver–ESIC
169 Federal Building
1961 Stout Street
Denver, CO 80294
303–844–4169; FTS 564–4169

Los Angeles–ESIC
Federal Building, Rm. 7638
300 North Los Angeles Street
Los Angeles, CA 90012
213–894–2850; FTS 798–2850

Menlo Park–ESIC
Building 3, MS 532, Rm. 3128
345 Middlefield Road
Menlo Park, CA 94025
415–329–4309; FTS 459–4309

Rolla–ESIC
1400 Independence Road
Rolla, MO 65401
314–341–0851; FTS 759–0851

Salt Lake City–ESIC
8105 Federal Building
125 South State Street
Salt Lake City, UT 84138
801–524–5652; FTS 588–5652

San Francisco–ESIC
504 Custom House
555 Battery Street
San Francisco, CA 94111
415–705–1010; FTS 465–1010

Sioux Falls–ESIC
EROS Data Center
Sioux Falls, SD 57198
605–594–6151; FTS 753–7151

Spokane–ESIC
678 U.S. Courthouse
West 920 Riverside Avenue
Spokane, WA 99201
509–353–2524; FTS 439–2524

Stennis Space Center–ESIC
Building 3101
Stennis Space Center, MS 39529
601–688–3544; FTS 494–3544

appendix

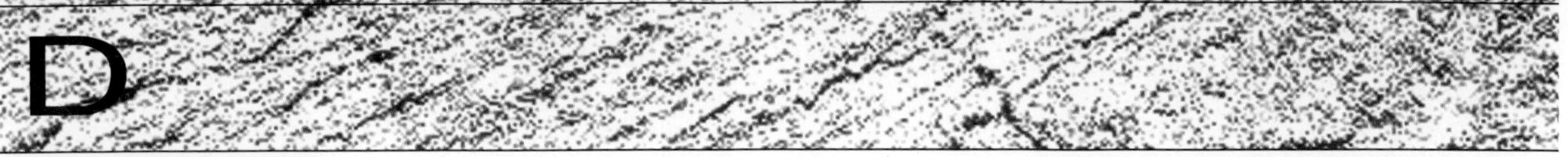

D

# Journals with Environmental Concerns

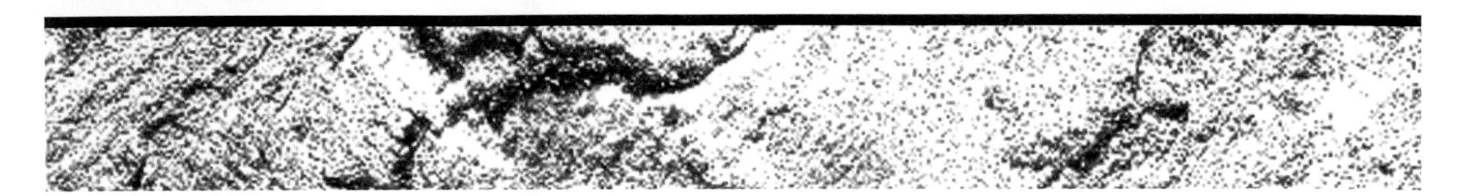

A very large number of professional journals publish articles that deal with environmental problems, and the number is increasing rapidly as concern over environmental problems mounts. Some journals have catchy names such as *Sludge;* or *Garbage,* but most of the titles are more drab. The list below includes many frequently cited monthly, bimonthly, or quarterly periodicals; references in the articles they contain will quickly lead you to other important journals. In addition to the periodicals listed, organizations such as the U.S. Geological Survey, the Environmental Protection Agency, and various arms of state governments issue at irregular intervals publications that discuss environmental problems.

*Coastal Management Journal*
*Environment*
*Environmental Geochemistry and Health*
*Environmental Impact Assessment Bulletin*
*Environmental Protection*
*Environmental Quality and Water Sciences*
*Environmental Science and Technology*
*Journal of the Air & Waste Management Association*
*Journal of Coastal Research*
*Journal of Contaminant Hydrology*
*Journal of Environmental Economics and Management*
*Journal of Environmental Management*
*Journal of Environmental Quality*
*Journal of Environmental Science and Health*
*Journal of Environmental Systems*
*Journal of Ocean and Shoreline Management*
*Journal of Soil and Water Conservation*
*Journal of the Water Pollution Control Federation*
*Natural Hazards Observer*
*Natural Resources Journal*
*Oceanus*
*Resources, Conservation, and Recycling*
*Water Research*
*Water Resources Bulletin*
*Water Resources Research*

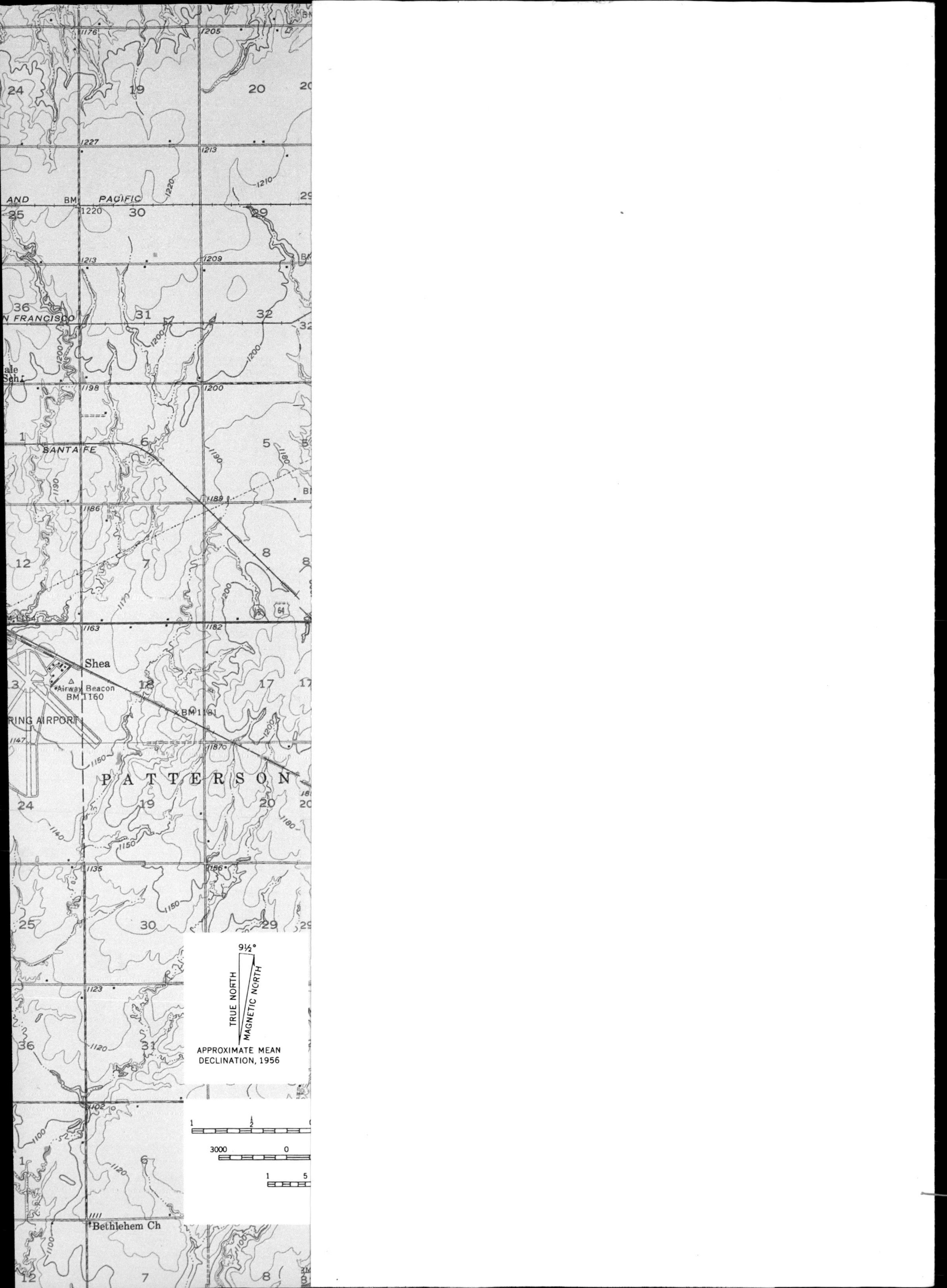

PACIFIC
N FRANCISCO
SANTA FE
Shea
Airway Beacon
BM 1160
RING AIRPORT
BM 1181
P A T T E R S O N
Bethlehem Ch
TRUE NORTH
MAGNETIC NORTH
9½°
APPROXIMATE MEAN
DECLINATION, 1956